Leckie×Leckie

D0317366

Intermediate 2
Chemistry

Archie Gibb ✕ David Hawley

Contents

Unit 1: Building blocks

Substances
Elements	4
Some families in the Periodic Table	6
Compounds and mixtures	8

Reaction rates
Following the course of a reaction	10
Factors affecting the rate of reaction	12
Catalysts	14

The structure of the atom
Atoms	16

Bonding, structure and properties
Bonding	18
Structure	20
More structures	22
Properties	24
Electrolysis and ion migration	26

Chemical symbolism
Formulae: two-element compounds and group ions	28
Formulae: using Roman numerals and brackets	30
Balanced equations	32

The mole
Calculations involving number of moles and using balanced equations	34

Prescribed practical activities
PPA 1: The effect of concentration changes on reaction rate	36
PPA 2: The effect of temperature changes on reaction rate	38
PPA 3: Electrolysis	40

Unit 2: Carbon compounds

Fuels
Combustion	42
Fractional distillation	44

Nomenclature and structural formulae
Hydrocarbons	46
More hydrocarbons	48
Isomers	50
Alkanols and alkanoic acids	52
Esters	54

Reactions of carbon compounds
Addition reactions 56
Cracking 58
Ethanol 60
Making and breaking esters 62

Plastics and synthetic fibres
Uses of plastics and synthetic fibres 64
Addition polymerisation 66
Condensation polymerisation 68

Natural products
Carbohydrates 70
Proteins 72
Fats and oils 74

Prescribed practical activities
PPA 1: Testing for unsaturation 76
PPA 2: Cracking 78
PPA 3: Hydrolysis of starch 80

Unit 3: Acids, bases and metals

Acids and bases
The pH scale 82
Concentration 84
Strong and weak acids and bases 86

Salt preparation
Reactions of acids 88
Volumetric titrations 90
Naming salts 92
Precipitation and formulae of ions 94
Ionic equations 96

Metals
The electrochemical series 98
Redox reactions 100
Reactions of metals 102
Metal ores 104
Corrosion 106

Prescribed practical activities
PPA 1: Preparation of a salt 110
PPA 2: Factors affecting voltage 112
PPA 3: Reaction of metals with oxygen 114

Index 116

Elements

What is an element?

Everything in the world is made from just over 100 **elements**. Every element is made up of very small particles called **atoms.** An element is a substance made up of only one type of atom. Each chemical element has a name and a symbol. The names and symbols of the different elements are given in the **Periodic Table** on page 8 of your Data Booklet: You should try to learn these.

- **Magnesium** is an element and it has the symbol, Mg. All magnesium atoms are the same.
- **Oxygen** is another element. It has the symbol, O. All oxygen atoms are the same but are different from magnesium atoms.

Top Tip

Remember that the Periodic Table in your Data Booklet has the names and symbols of the elements

Solids, liquids or gases?

Most chemical elements are **solid** at room temperature. Some elements exist as **gases** but only two elements, mercury (Hg) and bromine (Br) are **liquid** at room temperature. The state at room temperature (approximately 25°C) can be determined by looking at the melting and boiling points of the elements on page 3 of your Data Booklet.

- If 25°C is below the melting point of the element, then the element is a solid at room temperature.
- If 25°C is between the melting and boiling points of the element, then the element is a liquid at room temperature.
- If 25°C is above the boiling point of the element, then the element is a gas at room temperature.

Top Tip

Your Data Booklet is a source of very useful information. Make yourself familiar with its contents.

Metals or non-metals?

Approximately three-quarters of the elements in the Periodic Table are **metals**. Elements shown on the left-hand side of the dark zigzag line on the Periodic Table are metals, except for hydrogen, element number 1, which is a **non-metal**.

Naturally occurring or made by scientists?

Most elements are found naturally, either as the element or in compounds. The elements after uranium (number 92) have been made by scientists and do not occur naturally.

Group 1	Group 2												Group 3	Group 4	Group 5	Group 6	Group 7	Group 0
1 H																		2 He
3 Li	4 Be												5 B	6 C	7 N	8 O	9 F	10 Ne
11 Na	12 Mg						transition metals						13 Al	14 Si	15 P	16 S	17 Cl	18 Ar
19 K	20 Ca	21 Sc	22 Ti	23 V	24 Cr	25 Mn	26 Fe	27 Co	28 Ni	29 Cu	30 Zn		31 Ga	32 Ge	33 As	34 Se	35 Br	36 Kr
37 Rb	38 Sr	39 Y	40 Zr	41 Nb	42 Mo	43 Tc	44 Ru	45 Rh	46 Pd	47 Ag	48 Cd		49 In	50 Sn	51 Sb	52 Te	53 I	54 Xe
55 Cs	56 Ba	57 La	58–71	72 Hf	73 Ta	74 W	75 Re	76 Os	77 Ir	78 Pt	79 Au	80 Hg	81 Tl	82 Pb	83 Bi	84 Po	85 At	86 Rn
87 Fr	88 Ra	89 Ac	90–103															

Key

- Atomic number / Symbol
- naturally occurring solids
- gases
- liquid
- made by scientists

metals | non-metals

* at 28 atmospheres
† sublimes

57 La	58 Ce	59 Pr	60 Nd	61 Pm	62 Sm	63 Eu	64 Gd	65 Tb	66 Dy	67 Ho	68 Er	69 Tm	70 Yb	71 Lu
89 Ac	90 Th	91 Pa	92 U	93 Np	94 Pu	95 Am	96 Cm	97 Bk	98 Cf	99 Es	100 Fm	101 Md	102 No	103 Lr

Classification of elements

Chemists have classified elements by arranging them in the Periodic Table. A **group** is a column of elements in the Periodic Table, for example, fluorine, chlorine, bromine and iodine are elements in Group 7.

A **period** is a horizontal row of elements in the Periodic Table, for example, lithium, beryllium, boron and carbon are four elements in the second period.

Quick Test

1. Name and write the symbol for the only metal that is not solid at room temperature.
2. Which other element is liquid at room temperature?
3. Which three elements in Group 4 are metals?
4. Which element in Group 5 is a gas at room temperature?
5. In which group are all the elements gases?
6. Name two elements that are not naturally occurring and have been made by scientists.
7. Name an element in Group 7 that is solid at room temperature.
8. Which two elements in Group 7 are gases at room temperature?
9. Use the Periodic Table to find out what you can about sodium and xenon.

Answers 1. Mercury (Hg). **2.** Bromine (Br). **3.** Germanium (Ge), tin (Sn) and lead (Pb). **4.** Nitrogen. **5.** Group 0; the noble gases. **6.** Any with atomic number greater than 92. **7.** Iodine. **8.** Fluorine and chlorine. **9.** Sodium has the symbol Na. It is a naturally occurring metal that is solid at room temperature. Xenon has the symbol Xe. It is a naturally occurring gas. It is a non-metal.

Some families in the Periodic Table

The Periodic Table

The vertical columns in the Periodic Table are called **groups**. All the **elements** in any one **group** have similar chemical properties (they react in the same way during chemical reactions).

Groups in the Periodic Table include the alkali metals (Group 1), the halogens (Group 7) and the noble gases (Group 0).

Top Tip
Elements in the same group have similar chemical properties.

Group 1: The alkali metals

Examples: **lithium**, **sodium** and **potassium**.

The alkali metals react vigorously with water, releasing hydrogen gas and forming alkaline solutions.

Their reactivity increases down the group so sodium is more reactive than lithium but less reactive than potassium.

1	2
H 1	
Li 3	
Na 11	
K 19	
Rb 37	
Cs 55	
Fr 87	

Top Tip
The alkali metals are very reactive metals and must be stored under oil to stop air or water getting to them and causing a reaction.

The transition metals

The transition metals are found between Groups 2 and 3 in the Periodic Table. Examples include **iron**, **nickel** and **copper**.

- They are much less reactive than the metals in Group 1 and do not react quickly with oxygen or water.
- Transition metals are widely used. For example, iron is often used as a structural material in buildings, cars and bridges. Copper is a good conductor of both heat and electricity, and it is often used for electrical cables.
- Many transition metals and their compounds can act as catalysts. Iron and platinum are widely used in this way.

Group 7: The halogens

- The halogens are the most reactive non-metal elements.
- Down the group each element becomes less reactive, so fluorine is the most reactive and iodine is the least reactive.

Fluorine
- Fluorine is a very poisonous, pale yellow gas.

Chlorine
- Chlorine is a poisonous, pale green gas.
- It is used in water purification and bleaching.

Bromine
- Bromine is a poisonous, dense, brown liquid.

Iodine
- Iodine is a dark grey, crystalline solid.
- Iodine solution is used as an antiseptic in hospital operations.

3	4	5	6	7	0
				F 9	
				Cl 17	
				Br 35	
				I 53	
				At 85	

Top Tip
The halogens are the most reactive non-metals.

Group 0: The noble gases

The noble gases are the elements helium, neon, argon, krypton, xenon and radon.
- The noble gases are colourless gases and are very unreactive.

Argon
- Argon makes up about 1%, by volume, of dry air.
- It is used in light bulbs. Surrounding the hot filament with inert argon stops the filament from burning away.

Helium
- Helium is used in balloons and in airships, because it is less dense than air (and not flammable like hydrogen).

Krypton
- Krypton is used in lasers.

Neon
- Neon is used in electrical discharge tubes in advertising signs.

Top Tip
Remember that the noble gases are the least reactive of all the elements.

Quick Test

1. What are the vertical columns in the Periodic Table called?
2. Why are alkali metals stored under oil?
3. Which gas is released when alkali metals react with water?
4. Give two uses of transition metals.
5. Which is the most reactive halogen?
6. What is chlorine used for?
7. Name and give the symbols for the first three noble gases.
8. What volume of argon will be present in 1 litre of dry air?
9. What is argon used for?

Answers 1. Groups. **2.** Because they are very reactive and oil stops air and water getting to them. **3.** Hydrogen. **4.** Examples include: iron (structural material in buildings, cars and bridges), copper (electrical cables). **5.** Fluorine. **6.** Bleach, disinfectants. **7.** Helium (He), Neon (Ne) and Argon (Ar). **8.** 10 cm^3. **9.** It is the gas inside light bulbs.

Compounds and mixtures

Compounds

Compounds are formed when different elements react together. A compound contains two or more different elements chemically joined together. Sodium chloride is a compound formed when the elements sodium and chlorine react together. The word equation that describes this reaction is:

sodium + chlorine ⟶ sodium chloride
elements   **compound**

- Compounds with names ending in -ide contain the two elements indicated in the name.
  Sodium chloride, NaCl, contains the two elements sodium and chlorine.
- Compounds with names ending in -ate or -ite also contain the element oxygen.
  Sodium carbonate, Na_2CO_3, contains the elements sodium, carbon and oxygen.

Top Tip

Remember: if the name of the compound ends in -ate or -ite then the compound contains oxygen in addition to the other named elements.

Mixtures

When two substances are mixed together but don't actually react with each other (they do not chemically join) we say that a mixture has been formed. Mixtures are easy to separate. Common mixtures include crude oil and air. Crude oil is a mixture of different hydrocarbons. Air is a mixture of nitrogen (78%), oxygen (21%) and other gases. The test for oxygen is that it relights a glowing splint. There is not enough oxygen in the air for air to relight a glowing splint, but there is enough to keep it burning once lit.

Solutions

When a solute dissolves in a solvent a solution is formed. A solution is a mixture, because the solute has not reacted with the solvent when it dissolves.

- A substance that dissolves in a liquid is soluble.
- A substance that does not dissolve is insoluble.
- When no more of the substance will dissolve, a saturated solution has been formed.
- A dilute solution contains a lower concentration of solute than a concentrated solution.
- A concentrated solution can be diluted by adding water.

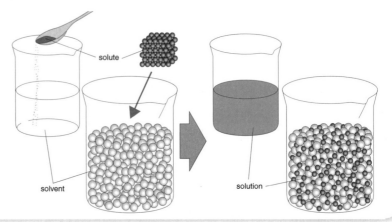

solute

solvent

solution

Chemical reactions

When a chemical reaction takes place a new substance is always formed. The substances at the start of the reaction are known as the reactants and the new substances formed are known as the products.

Identifying chemical reactions

Three things that can indicate a chemical reaction has taken place include:

- a colour change
- a precipitate forming
- a gas being given off.

Top Tip
Remember that a precipitate is the solid formed when two liquids react together.

Exothermic and endothermic reactions

Sometimes you cannot see any change when a chemical reaction takes place, but there may be an energy change such as a change in temperature. An example of this would be adding a dilute acid to a dilute alkali in a test tube. There would be no colour change, no gas given off and no solid formed. However, heat would be given out in the reaction, and the contents of the test tube would show a temperature rise. A reaction, like this, in which heat is given out is said to be exothermic. If the temperature drops, the reaction is said to be endothermic.

Reaction profiles for exothermic and endothermic reactions are shown below. In each case A is the reactant and B is the product of the reaction.

Exothermic

Exothermic reactions release energy to the surroundings and the products have less energy than the reactants.

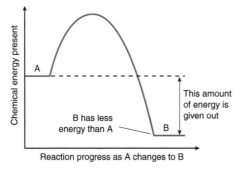

Endothermic

Endothermic reactions take in energy from the surroundings and the products have more chemical energy than the reactants.

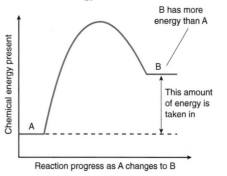

Quick Test

1. Which elements are present in: magnesium oxide, zinc carbonate, sodium sulphite, sodium chloride, sodium chlorate?

2. What is the difference between a compound and a mixture?

3. Which of the changes below are not chemical reactions:
   a) melting ice   b) frying an egg   c) adding water to orange juice
   d) iron rusting   e) magnesium burning?

4. What does exothermic mean?

Following the course of a reaction

Measuring changes

The rate of a chemical reaction is a measure of how fast the reactants are being used up and how fast the products are being made. The rate can be determined by measuring:

- changes in the concentration of the reactants or products
- changes in the mass of the reactants or products
- changes in the volume of the reactants or products.

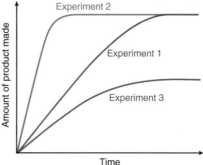

For example, consider the reaction between calcium carbonate and hydrochloric acid to produce calcium chloride solution, water and carbon dioxide gas. As the reaction proceeds, the hydrochloric acid is getting used up and so its concentration is decreasing. Since carbon dioxide gas is also given off in the reaction the mass of the reactants is decreasing and the volume of the products is increasing.

The graph shows the amount of product made in three different experiments starting from the same reactants each time.

The graph is steepest at the start of the reaction for all three experiments because this is where the reaction has most reactants. It then becomes less steep as the reactants get used up. When the graph becomes level, the reaction has finished. The graph shows that Experiment 2 is faster than Experiment 1.

Experiments 1 and 2 make the same amount of product so they both had the same amounts of reactants at the start of the reaction. In Experiment 3 only half as much product was made. This shows that there were less reactants at the start of Experiment 3.

Sometimes we can measure the time taken for an event to happen in a chemical reaction. For example, we can measure the time taken for a certain mass of magnesium to react with an acid or the time it takes for a certain colour change to take place. The rate of reaction can be taken to be the reciprocal of the time taken (1/t).

- The longer the time taken then the slower is the reaction.
- The shorter the time taken then the faster is the reaction.

If the time is measured in seconds (s) then the units for the rate of reaction will be s^{-1}.

Top Tip
When the mass of the reaction mixture decreases during a chemical reaction this is because a gas is given off during the reaction.

Calculating the average rate of reaction

The graph below was drawn using the results of an experiment in which an acid was used as one of the reactants. The concentration of the acid was measured at regular time intervals as the reaction progressed. The units of concentration are mol l^{-1}. As you can see the concentration of the acid is highest at the start of the reaction and decreases as the reaction proceeds.

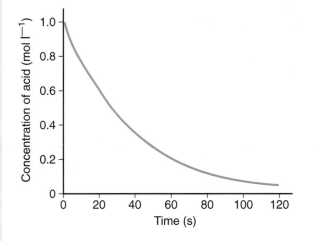

Top Tip
When analysing graphs remember:
- the steeper the slope the faster the reaction
- the reaction is over when the graph levels out.

A graph like this can be used to calculate the average rate of a reaction or stage in a reaction. For example, if we want to calculate the average rate of the reaction over the first 20 seconds:

Average rate of reaction over first 20 seconds $= \dfrac{\text{Concentration of acid at 0s} - \text{Concentration of acid at 20s}}{\text{Time interval}}$

$$= \frac{1.0 - 0.6}{20} = \frac{0.4}{20} = 0.020 \text{ mol } l^{-1} \text{ s}^{-1}$$

(Since the rate is a measure of concentration changes over time then the unit of rate is mol l^{-1} s^{-1}.)

Quick Test

1. When calcium carbonate is mixed with hydrochloric acid in a conical flask which is sitting on a balance the mass decreases. What must be happening to cause this decrease in mass?

2. Using the graph above calculate the average rate of the reaction:

   a) over the period 20–60 seconds

   b) over the period 0–60 seconds.

Answers 1. There must be a gas given off in the reaction. (In this case the gas is carbon dioxide, CO_2.) **2. a)** 0.010 mol l^{-1} s^{-1}. **b)** 0.013 mol l^{-1} s^{-1}.

Factors affecting the rate of reaction

Collision theory

Everything in the world is made up of tiny particles. Later on in this unit we will look more closely at these particles and their structure. For a chemical reaction to take place the reacting particles have to collide with each other. This is the basis of the collision theory.

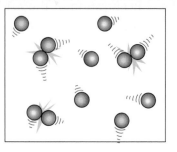

Particle size/surface area

The greater the surface area, the more chance of collisions occurring, so the faster the rate of reaction. To increase the surface area, a large lump should be broken down into smaller lumps or crushed into a powder. Powder has a much greater overall surface area than lumps of the same mass. For example, calcium carbonate powder reacts faster with dilute acid than does a large lump of calcium carbonate.

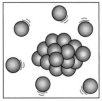 small surface area (larger pieces)

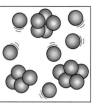 large surface area (smaller pieces)

The greater the surface area, the more collisions there will be and so the reaction rate will be greater.

Concentration

The greater the concentration, the more reactant particles there are moving in the solution. There will be more collisions and so the reaction rate is increased. For example, 2.0 mol l^{-1} hydrochloric acid reacts more quickly with calcium carbonate than does 1.0 mol l^{-1} hydrochloric acid.

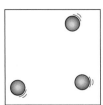 low concentration of particles

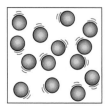 high concentration of particles

Top Tip
Remember that the more collisions there are then the faster the reaction will be.

The greater the concentration then the more particles in a given volume and so there will be more collisions and the reaction rate will be greater.

Temperature

If the temperature is increased, the particles have more energy and so move quicker. Increasing the temperature increases the rate of reaction because the particles collide more often and with more energy so there are more successful collisions, and the reaction rate increases. Very often a good way to start a chemical reaction or to get it to go faster is to heat the reactants.

 A low temperature slows down reaction rate.

 A high temperature speeds up reaction rate.

Increasing the temperature speeds up a chemical reaction by giving the reacting particles more energy.

Adding a catalyst

A catalyst is a substance that increases the rate of a chemical reaction. Catalysts are not themselves used up by the reaction. They are still present at the end of the reaction. Catalysts are usually transition metals or compounds containing transition metals. Adding copper sulphate speeds up the reaction between zinc and sulphuric acid. Copper sulphate is therefore a catalyst for this reaction. If a catalyst is present, the reacting particles can collide more successfully with less energy and so the reaction can take place at a lower temperature.

Top Tip

You should know four ways of speeding up chemical reactions:
- increasing the surface area
- increasing the concentration
- increasing the temperature
- adding a catalyst.

Quick Test

1. What happens to the rate of a reaction if the temperature is decreased?
2. What happens to the rate of a reaction if the concentration of the reactants is increased?
3. What happens to the rate of a reaction if the surface area of a reactant is increased?
4. What does a catalyst do to the rate of reaction?
5. Why can catalysts be reused?
6. Use the collision theory to explain why powders react faster than lumps.
7. Give four ways in which you could increase the rate of a chemical reaction.
8. Give three ways in which you could decrease the rate of a chemical reaction.

Answers 1. Decreases. **2.** Increases. **3.** Increases. **4.** A catalyst increases the rate of reaction. **5.** Catalysts are not used up by the reaction and so can be reused. **6.** Powders have a greater surface area and so more collisions will take place. The more collisions then the faster the reaction. **7.** Increase temperature; increase concentration; increase surface area/decrease particle size; add a catalyst. **8.** Decrease temperature; decrease concentration; decrease surface area.

Catalysts

Heterogeneous or homogeneous

All catalysts can be classified as either **homogeneous** or **heterogeneous**.

A **homogeneous** catalyst is in the **same state** as the reactants. For example, if the reactants are in aqueous solution then a homogeneous catalyst would have to be in aqueous solution.

A **heterogeneous** catalyst is in a **different state** to the reactants. For example, if the reactants are gases and the catalyst is a solid, then it is a heterogeneous catalyst.

Top Tip
Make sure you know the difference between homogeneous and heterogeneous catalysts.

Uses of catalysts

Since reactions that are catalysed will work at lower temperatures than the same uncatalysed reaction then catalysts are used in many industrial processes. This keeps energy costs down, so means that the final product can be made more cheaply making the industrial process more profitable.

Some examples of catalysts used in the chemical industry are shown in the table below.

Name of Process	Reactants	Product	Catalyst
Haber	Nitrogen and hydrogen gases	Ammonia	Iron
Making margarine	Vegetable oils and hydrogen	Margarine	Nickel
Ostwald	Ammonia and oxygen	Nitric acid	Platinum

How do heterogeneous catalysts work?

Heterogeneous catalysts have **active sites** on their surface. The reacting particles are held in place on these active sites so that other reacting particles collide with them. The reactant particles are held in these active sites by the process known as **adsorption**. After a successful collision has taken place and the products have been formed, the products leave the active site. The vacant active site then becomes available for another reactant particle to become adsorbed and the process continues.

Catalyst poisoning

If impurities are present along with the reactants then particles of the impurity may become adsorbed onto the active sites. The impurity blocks the active site stopping reactant particles being adsorbed and this is known as **poisoning**. In some cases the poisoned catalyst can be **regenerated** simply by burning off the impurity which has been adsorbed onto its surface. If it is not possible to regenerate the catalyst, then the catalyst has to be **renewed**. Since catalysts are often expensive metals such as platinum, this can be very costly. It is very important, in industry, to remove impurities from reactants before they reach the catalyst.

The catalyst in the catalytic converter fitted into the exhaust system of cars is a mixture of three metals, platinum, palladium and rhodium. This is a heterogeneous catalyst since it is a solid and the reactants are gases. Poisonous carbon monoxide and oxides of nitrogen are converted to carbon dioxide and nitrogen. Lead-free petrol must be used since lead would poison the catalyst.

Top Tip
Impurities adsorbed onto the active sites of the catalyst cause the catalyst to become **poisoned.**

Enzymes

Enzymes are **biological catalysts**. This means that enzymes catalyse chemical reactions occurring in the living cells of animals and plants. For example, the different reactions taking place in our digestive system are catalysed by different enzymes. Enzymes are specific. This means they can only catalyse one reaction. The enzyme that catalyses the breakdown of the carbohydrate starch in our digestive system could not catalyse the breakdown of protein.

Biological washing powders contain enzymes that are said to 'digest' stains on clothes. Enzymes are also used to make soft centres in chocolates, to tenderise beef and in the manufacture of yoghurt and cheese.

The enzyme **zymase** is used to catalyse the reaction of glucose into ethanol (alcohol). This is very important in the brewing and whisky industries.

Top Tip
Remember that enzymes catalyse chemical reactions taking place in plants and animals.

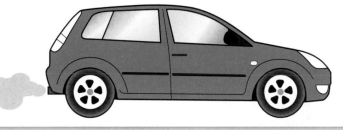

Quick Test

1. What is meant by the following words:
   a) catalyst   b) homogeneous   c) heterogeneous   d) enzyme
2. Write down an example of a heterogeneous catalyst.
3. What happens to a catalyst when it is poisoned?
4. What are the products when carbon monoxide and oxides of nitrogen pass through a catalytic converter?

Atoms

Sub-atomic particles

Every element is made up of very small particles called **atoms**. Atoms have a positively charged **nucleus**, containing positively charged **protons** and neutral **neutrons**. Negatively charged **electrons** move around outside the nucleus. An atom is neutral because the number of protons is equal to the number of electrons.

An atom has a nucleus surrounded by shells of electrons.

The **electrons** are found in shells around the **nucleus**.

The **nucleus** is found at the centre of the **atom** and contains **neutrons** and **protons**.

- Protons have a positive charge and a mass of 1.
- Neutrons have no charge and also have a mass of 1.
- Electrons have a negative charge and virtually no mass (we say electrons have zero mass).

Top Tip
Make sure you know the charge, mass and location of the three types of particles in the atom.

Important numbers

Atoms of different elements have a different number of protons. The number of protons in an atom is known as the **atomic number**.

The sum of the protons and neutrons in an atom is known as the **mass number**. Consider the information below about sodium, in which the sodium atom is represented in a form which we call its nuclide notation.

23   The mass number is the number of protons added to the number of neutrons
 Na Symbol of element
11   The atomic number is the number of protons only

- Sodium has an atomic number of 11, so it has 11 protons.
- The sodium atom has no overall charge so it must also have 11 electrons.
- The number of neutrons is given by the mass number minus the atomic number. Sodium therefore has 23 – 11 = 12 neutrons.

Elements are arranged in the Periodic Table in terms of their atomic numbers and chemical properties. The electron arrangements of elements are given on page 1 of your Data Booklet. Elements in the same group of the Periodic Table have similar chemical properties. We can see from their electron arrangements that all elements within a group have the same number of outer electrons and so we can conclude that elements with the same number of outer electrons have similar chemical properties.

Top Tip
The number of neutrons = mass number – atomic number.

Isotopes

Isotopes of an element have the same number of protons but a different number of neutrons. So isotopes have the same atomic number but a different mass number.

Chlorine has two common isotopes:

35	17 protons
Cl	17 electrons
17	18 neutrons

37	17 protons
Cl	17 electrons
17	20 neutrons

Top Tip
You must know what is meant by "isotopes".

The isotopes will react chemically in the same way because they have identical numbers of electrons.

Most elements exist as a mixture of different isotopes and the **relative atomic mass** (RAM) is the **average atomic mass** and so is rarely a whole number. The RAM is used in calculations. The approximate RAM of selected elements is given on page 4 of your Data Booklet and these have been rounded to the nearest 0.5. You can see that the RAM of chlorine is given as 35.5. Chlorine has 2 isotopes, ^{35}Cl and ^{37}Cl and the average mass, 35.5, is closer to 35 than 37. This tells us that ^{35}Cl is more abundant than ^{37}Cl.

Quick Test

1. Copy and complete the table below

Particle	Mass	Charge	Location
Proton			
Neutron			
Electron			

2. What is the same about the atoms of two isotopes of an element?

3. What is different about the atoms of two isotopes of an element?

4. Why do isotopes of an element react in the same way?

5. What is the same for all elements in the same group in the Periodic Table?

6. Copy and complete the table below:

Element	Atomic Number	Mass Number	Number of protons	Number of Neutrons	Electron Arrangement
Sodium	11	23			
Calcium		42			
Bromine		79			

Answers 1. Proton 1 +1 In the nucleus; Neutron 1 0 In the nucleus; Electron 0 −1 Outside the nucleus. **2.** Atomic number/number of protons or number of electrons. **3.** Mass number/number of neutrons. **4.** They have the same number of electrons in the outer shell. **5.** The number of outer electrons. **6.** Sodium 11 23 11 12 2,8,1; Calcium 20 42 20 22 2,8,8,2; Bromine 35 79 35 44 2,8,18,7.

Bonding

Covalent bonding and polar covalent bonding

Atoms can be held together by **bonds**. When atoms form bonds, they can achieve a stable electron arrangement (usually eight outer electrons). To achieve a stable electron arrangement atoms can lose, gain or share electrons. There are different types of bonds that hold atoms together.

In **covalent bonding** the atoms share pairs of electrons. Covalent bonding usually occurs between non-metal atoms. A covalent bond happens when the positive nuclei from two different atoms are held together by their common attraction for the shared pair of electrons held between them. The covalent bond is really this shared pair of electrons holding the positive nuclei together. Covalent bonds are strong bonds.

When the two atoms that are sharing the bonding electrons have an equal pull on these shared electrons, then there is an equal sharing of the electrons and the bonding is **non-polar covalent bonding**. For example, non-polar covalent bonding will be found in the **diatomic** elements such as hydrogen (H_2) and chlorine (Cl_2), and, since carbon and hydrogen have a similar pull on the shared electrons, methane (CH_4) also has non-polar covalent bonding between its atoms.

When one element has a stronger pull on the shared electrons than the other element, then the atom with the greater pull takes on a slightly negative charge and the other atom takes on a slightly positive charge. This is known as **polar covalent bonding**.

Water, H_2O, has polar covalent bonding. The oxygen atom takes on a slight negative charge and both hydrogen atoms take on slight positive charges. This is because oxygen has the greater pull on the shared electrons. The water molecule can be represented as shown below:

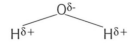

(The symbol for slightly is δ, so $\delta+$ means slightly positive and $\delta-$ is slightly negative.)

Top Tip
Remember covalent and polar covalent bonding both involve the sharing pairs of electrons.

Ionic bonding

Ionic bonding involves the **transfer** of electrons from one atom to another. When this happens the atoms involved end up with full outer shells of electrons, similar to atoms of the noble gases (usually eight outer electrons). Since one atom has lost one or more electrons and the other atom has gained them, these atoms now have positive or negative charges and are now called **ions**. When electrons are lost or gained then the number of protons and the number of electrons is no longer balanced, so an ion is an atom, or a small group of atoms with a charge. **Ionic bonding** is the **electrostatic** force of attraction between positively charged ions and negatively charged ions. Ionic bonds are strong bonds.

Ionic compounds are usually formed when metals combine with non-metals. **Positive ions** are formed when metal atoms lose electrons and **negative ions** are formed when non-metal atoms gain electrons. For example, a sodium atom has electron arrangement 2,8,1. When sodium reacts to form an ionic compound, each sodium atom loses its outer electron to form a sodium ion which has the more stable electron arrangement 2,8. After the reaction, a sodium ion will have 11 positive protons but only 10 negative electrons, so will have an overall 1+ charge. The sodium ion is therefore written as Na^+.

The table below shows the ions formed by the first 20 elements.

Top Tip
Metal ions are positive, and non-metal ions are usually negative.

H+ hydrogen								Group 0
								None helium
Group 1	**Group 2**		**Group 3**	**Group 4**	**Group 5**	**Group 6**	**Group 7**	
Li+ lithium	Be²⁺ beryllium		None boron	None carbon	N³⁻ nitride	O²⁻ oxide	F⁻ fluoride	None neon
Na+ sodium	Mg²⁺ magnesium		Al³⁺ aluminium	None silicon	P³⁻ phosphide	S²⁻ sulphide	Cl⁻ chloride	None argon
K+ potassium	Ca²⁺ calcium	transition metals						

Top Tip
Names of negative non-metal ions end in -ide e.g. oxide, fluoride, etc.

Metallic bonding

Metallic bonding is the type of bonding found in metals. Metal atoms hold on to their outer electrons very weakly, and these outer electrons usually leave the atoms so that positive ions are formed. The outer electrons are now free to move between the positive ions and stop these ions repelling each other. Since these outer electrons are not stuck in a particular locality they are said to be **delocalised**. Metallic bonding is the electrostatic force of attraction between positively charged metal ions and delocalised electrons.

Quick Test

1. What is the main difference between non-polar covalent bonding and polar covalent bonding?

2. What is the main difference between covalent bonding and ionic bonding?

3. What type of bonding is likely to be found in:
   a) sodium chloride
   b) sodium
   c) chlorine
   d) hydrogen
   e) ammonia, NH₃?

Answers 1. In polar covalent bonding the electrons are shared unequally. **2.** In covalent bonding the electrons are shared, but in ionic bonding there is a transfer of electrons. **3. a)** ionic **b)** metallic **c)** non-polar covalent **d)** non-polar covalent **e)** polar covalent.

Structure

Structures of covalent molecular substances

Atoms that share pairs of electrons form **molecules**. A molecule is a group of atoms held together by covalent bonds. These can be single, double or even triple bonds.

A **diatomic** molecule is a molecule containing only two atoms. An example is hydrogen chloride, HCl, in which each molecule contains one hydrogen atom joined to one chlorine atom. Another example is carbon monoxide, CO.

Diagrams can be used to show how the outer electrons are shared to form the covalent bonds in a molecule.

In the following diagrams, the outer electrons from one atom are represented by dots and the outer electrons from the other atom are represented by crosses.

Top Tip
A **molecule** is a group of atoms held together by covalent bonds.

Hydrogen, H₂

Both hydrogen atoms have only one electron, but by forming a single covalent bond, both can have a full outer shell.

This can also be shown as **H — H**

Hydrogen chloride, HCl

The hydrogen and the chlorine atoms both need **one** more electron. They form a single covalent bond so both have a full outer shell.

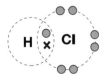

Molecules such as H₂ and HCl are **diatomic** (there are two atoms in a molecule). **Elements** which exist as **diatomic molecules** are H₂, N₂, O₂, F₂, Cl₂, Br₂ and I₂.

Methane, CH₄

The carbon has four outer electrons so needs **four** more for a full outer shell. The carbon forms four single covalent bonds to the hydrogen atoms, so all the atoms now have a full outer shell of electrons.

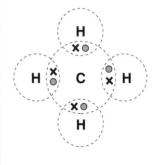

This can also be shown as

Ammonia, NH₃

The nitrogen atom has five outer electrons so needs three more. Nitrogen forms three single covalent bonds to hydrogen atoms.

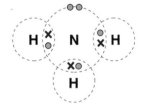

This can also be shown as

Top Tip
You need to be able to draw the shapes of water and methane molecules.

Water, H₂O

The oxygen has six outer electrons so needs two more. The oxygen forms two single covalent bonds with the two hydrogen atoms to give it a full outer shell.

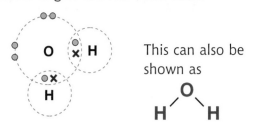

This can also be shown as

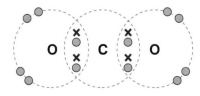

Oxygen, O₂

Both oxygen atoms have six outer electrons so both need two more. The oxygen atoms form one double covalent bond so that both have a full outer shell.

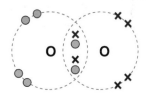

This can also be shown as **O = O**

Carbon dioxide, CO₂

The carbon has four outer electrons, so needs four more. It forms double covalent bonds with two oxygen atoms, so that all the atoms now have a full outer shell of electrons.

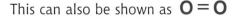

This can also be shown as **O = C = O**

The covalent substances shown above consist of individual molecules. The covalent bonds inside each molecule are strong bonds, but the bonds that can exist between molecules are much weaker. It is difficult to break the covalent bonds inside the molecules, but easy to overcome the bonds between the molecules. This explains why substances that consist of covalent molecules are usually gases or liquids at room temperature.

The formula for a covalent substance tells you the actual number of atoms of each element in a molecule. For example, the formula for water is H_2O. This means that one molecule of water contains one oxygen atom joined to two hydrogen atoms.

The structure of ethanol is:

```
    H   H
    |   |
H — C — C — O — H
    |   |
    H   H
```

You can see from the structure that one molecule of ethanol, C_2H_6O, consists of two carbon atoms, six hydrogen atoms and one oxygen atom.

Quick Test

1. What is meant by a molecule?

2. Draw the shape of a molecule of:
   a) water   **b)** methane   **c)** ammonia.

3. Draw a dot and cross diagram to show the bonding in methane.

4. Which seven elements exist as diatomic molecules?

5. Why is oxygen a gas at room temperature?

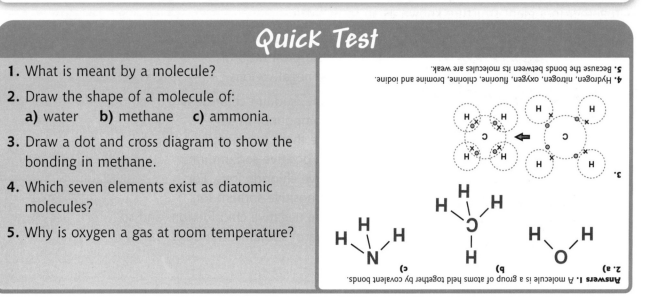

Answers 1. A molecule is a group of atoms held together by covalent bonds.
2. a) **b)** **c)**
3.
4. Hydrogen, nitrogen, oxygen, fluorine, chlorine, bromine and iodine.
5. Because the bonds between its molecules are weak.

More structures

Structures of covalent network substances

The previous two pages looked at covalent molecular substances. These consist of molecules containing small numbers of atoms. These substances are usually liquids or gases at room temperature. Some, such as iodine, sulphur and phosphorus are solids with low melting points.

Covalent substances that are solids with high melting points have much larger molecules. A **covalent network structure** consists of a giant **lattice** of covalently bonded atoms. Examples of covalent network substances are diamond, which is a pure form of carbon, and silicon dioxide.

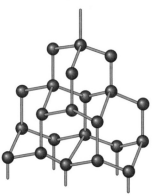

The structure of diamond

Covalent network structures contain many millions of atoms joined together by covalent bonds. The formula for a covalent network substance gives the simplest ratio of atoms of each element. Silicon dioxide is a covalent network substance and its formula is written as SiO_2. This means that for every one silicon atom in the structure there are two oxygen atoms.

Structures of ionic compounds

In ionic compounds the oppositely charged positive and negative ions are held together in a giant structure known as an **ionic lattice**.

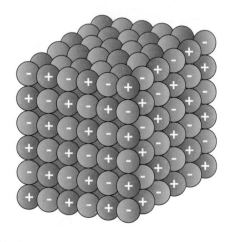

The formula for an ionic compound gives the simplest ratio of positive to negative ions.

For example, sodium chloride is an ionic compound. In its ionic lattice there are many millions of positive sodium ions and negative chloride ions. For every one positive sodium ion there is one negative chloride ion, and so the formula of sodium chloride is Na^+Cl^- or NaCl.

Top Tip

In an ionic lattice, each positive ion is surrounded by negative ions and vice versa. There are no molecules in ionic compounds.

Metallic structures

A metallic structure consists of a giant lattice of positively charged ions and delocalised outer electrons.

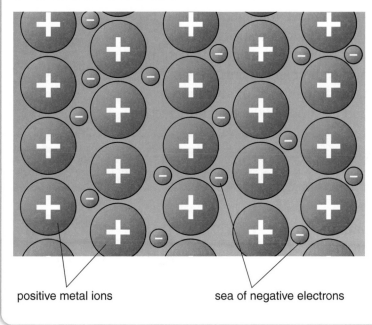

positive metal ions   sea of negative electrons

Quick Test

1. Name three chemical elements that are solid at room temperature but have covalent molecular structures.

2. Name two substances that have covalent network structures.

3. Silicon dioxide has the formula SiO_2. Does this mean that each molecule of silicon dioxide contains one silicon atom joined to two oxygen atoms?

4. Why is it incorrect to say 'molecules of sodium chloride'?

5. Describe briefly what is meant by
   a) a covalent network structure
   b) an ionic lattice
   c) a metallic structure.

Answers 1. Iodine, sulphur and phosphorus. **2.** Silicon dioxide and diamond. **3.** No, silicon dioxide has a covalent network structure and so does not have small SiO_2 molecules. It has a giant lattice structure and the simplest ratio of the atoms is one silicon atom for every two oxygen atoms. **4.** A molecule is a group of atoms held together by covalent bonds. Sodium chloride has ionic bonding not covalent bonding. Ionic compounds contain ions not molecules. **5. a)** A covalent network structure consists of a giant lattice of covalently bonded atoms. **b)** An ionic lattice is a giant structure containing oppositely charged ions. **c)** A metallic structure consists of a giant lattice of positively charged ions and delocalised outer electrons.

Properties

Properties of metals

Metals have delocalised electrons that are free to move throughout their structure. These 'free' electrons allow metals to conduct electricity. Metals conduct electricity in the solid state and when melted into a liquid. Apart from mercury, metals have high melting points and high boiling points. This is because a lot of energy is needed to break the strong forces of attraction between the positive ions and the delocalised electrons in the metallic structure.

Properties of graphite

Graphite is a form of pure carbon. It has a covalent network structure. The carbon atoms are in layers of hexagonal rings. Each carbon atom is covalently bonded to two other carbon atoms. Since carbon has four outer electrons, each carbon atom will have a spare outer electron not involved in covalent bonding. These spare electrons are delocalised electrons which allow graphite to conduct electricity.

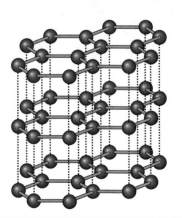

The structure of graphite

Properties of covalent substances

For a substance to conduct electricity it must have charged particles that are free to move.

Covalent substances are made up of molecules. Since molecules have no charge, covalent substances, except for graphite, do not conduct electricity. Discrete covalent molecular substances have low melting and boiling points and are often gases or liquids at room temperature. This is because there are only very weak forces of attraction between the individual molecules.

The diagram shows 3 discrete CO_2 molecules. Since only a small amount of energy is needed to overcome the weak forces, the molecules can move away from each other easily and so CO_2 is a gas at room temperature.

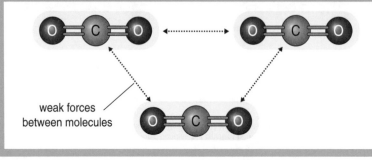

weak forces
between molecules

Top Tip
Melting and boiling points of some covalent and ionic compounds are given on page 6 of your Data Booklet. Solubility of some substances in water is given on page 5 of your Data Booklet.

Covalent network substances have high melting and boiling points. This is because it is necessary to break some of the strong covalent bonds inside the lattice to melt the solid into a liquid, and even more of these bonds have to be broken when a liquid changes to a gas at the boiling point.

Many covalent substances are insoluble or only slightly soluble in water. This is because the uncharged molecules are not attracted to the slight charges on the water molecules.

Covalent substances may be more soluble in other solvents such as petrol or white spirit.

Properties of ionic substances

When ionic compounds are solid, the ions are trapped within the ionic lattice and so cannot move, and therefore cannot conduct electricity. There are strong electrostatic forces of attraction between the oppositely charged ions. Breaking the lattice to allow the ions to move requires a great deal of energy, and so ionic compounds have very high melting and boiling points. In the **molten state**, when the ions are free to move, ionic compounds do conduct electricity.

Most ionic compounds are soluble in water. This is because the charged ions are attracted to the slightly charged atoms in the water molecules. When ionic compounds dissolve in water the ionic lattice breaks and the ions become free to move. In solution, when the ions are free to move, ionic compounds do conduct electricity.

Top Tip
Make sure you know that ionic compounds conduct electricity only when in solution or when molten. This is because the ions are free to move.

Quick Test

1. Why do covalent substances not conduct electricity?
2. Which covalent network substance does conduct electricity and why?
3. Explain which compound, silicon dioxide or carbon dioxide has the higher melting point.
4. Why are metals good conductors of electricity?
5. Why is oxygen a gas at room temperature?
6. In which states do ionic compounds conduct electricity?
7. Give two differences between metals conducting electricity and ionic compounds conducting electricity.

Answers 1. Covalent substances have no freely moving charged particles to conduct electricity. **2.** Graphite conducts electricity because it contains delocalised electrons. **3.** Silicon dioxide, as it has a covalent network structure. Carbon dioxide is covalent molecular. **4.** Metals are good conductors because they have delocalised or free electrons. **5.** Oxygen has covalent molecules with only very weak forces between the molecules. There is enough energy available at room temperature to overcome these weak forces. **6.** Molten and when in solution. **7.** Metals conduct in the solid state due to freely moving electrons carrying the current. Ionic substances only conduct when molten and when in solution when freely moving ions carry the current.

Electrolysis and ion migration

Coloured ions

Most ionic compounds are white when solid and dissolve in water to give colourless solutions. However, ions containing **transition metals** are often coloured. The table below shows some examples.

Row	Name of compound	Positive ions present	Negative ions present	Colour
1	sodium chloride	sodium	chloride	colourless
2	copper chloride	copper	chloride	blue
3	sodium chromate	sodium	chromate	yellow
4	potassium nitrate	potassium	nitrate	colourless
5	potassium permanganate	potassium	permanganate	purple
6	sodium sulphate	sodium	sulphate	colourless
7	nickel sulphate	nickel	sulphate	green
8	copper chromate	copper	chromate	green

- If we look at row 1 in the table we can see both sodium and chloride ions must be colourless.
- Remembering that chloride ions are colourless and looking at row 2, we can see that copper ions must be blue.
- If we now look at row 3, it should be obvious that chromate ions must be yellow.
- Copper chromate is coloured green because it contains blue copper ions and yellow chromate ions.

Ion migration

Ion migration experiments such as the one shown below demonstrate that the blue copper ions travel to the negative electrode and the yellow chromate ions travel to the positive electrode. This confirms that copper ions are positively charged and that chromate ions are negatively charged.

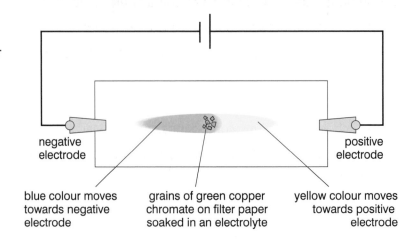

negative electrode

positive electrode

blue colour moves towards negative electrode

grains of green copper chromate on filter paper soaked in an electrolyte

yellow colour moves towards positive electrode

Electrolysis

A solution containing ions is an **electrolyte. Electrolysis** is when electricity is passed through an ionic solution. When an ionic compound is electrolysed, the **positive metal ions** move to the **negative electrode** where they pick up electrons and change into metal atoms. This **gain of electrons** is known as **reduction.**

At the same time, the **negative non-metal ions** travel to the **positive electrode** where they lose electrons and change into molecules. This **loss of electrons** is known as **oxidation.**

When ionic compounds conduct electricity the ionic compound can change or undergo **decomposition.**

For example, when copper chloride solution is electrolysed it changes into copper and chlorine. We say that the copper chloride has **decomposed** into copper and chlorine.

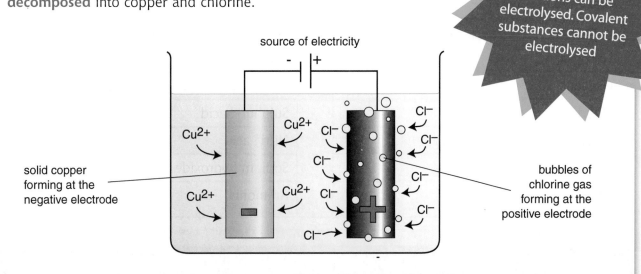

source of electricity

solid copper forming at the negative electrode

bubbles of chlorine gas forming at the positive electrode

Since positive and negative electrodes are needed in this type of experiment, a d.c. supply of electricity must be used if the products are to be identified.

Top Tip

Only molten ionic compounds and ionic solutions can be electrolysed. Covalent substances cannot be electrolysed

Quick Test

1. Why did the blue colour move towards the negative electrode in the ion-migration experiment?

2. What is meant by:
   a) electrolysis
   b) electrolyte?

3. Why do chloride ions travel to the positive electrode during the electrolysis of copper chloride solution?

Answers 1. The blue colour is due to the copper ions, and because they are positively charged they move to the negative electrode. **2. a)** Electrolysis is the passage of an electric current through an ionic solution or molten ionic compound. **b)** An electrolyte is a liquid containing ions. **3.** Chloride ions are negatively charged and so travel to the positive electrode during electrolysis.

Formulae: two-element compounds and group ions

Chemical formulae you must know

Water molecules contain two hydrogen atoms and one oxygen atom joined together and so water has the chemical formula H_2O. There are some formulae you should make an effort to learn including the elements that exist as diatomic molecules.

Name of diatomic element	Formula	Name of compound	Formula
Hydrogen	H_2	Water	H_2O
Nitrogen	N_2	Carbon dioxide	CO_2
Oxygen	O_2	Hydrochloric acid	HCl
Fluorine	F_2	Sulphuric acid	H_2SO_4
Chlorine	Cl_2	Nitric acid	HNO_3
Bromine	Br_2	Sodium hydroxide	$NaOH$
Iodine	I_2	Ammonia	NH_3

Formulae from names of compounds

Sometimes the name of the compound gives information about the formula of that compound. Names of these compounds have prefixes that give the number of atoms of certain elements in each molecule. (See table)

- Carbon **mono**xide contains one carbon atom joined to **one oxygen** atom, so it has the formula CO.
- **Di**nitrogen **tetr**oxide has the formula N_2O_4.

Prefix	Meaning
mono-	one
di-	two
tri-	three
tetra-	four
penta-	five

Valency

Valency is the combining power of an element. Elements in the same group of the Periodic Table have the same valency. Valencies of the elements in the different groups in the Periodic Table are shown in the table below. The noble gases have valency 0 as they do not usually combine with other elements.

Group in the Periodic Table	1	2	3	4	5	6	7	0
Valency	1	2	3	4	3	2	1	0

Using your knowledge of the Periodic Table you should be able to work out the valencies of different elements.

- Sodium is in Group 1 so Na will have valency = 1
- Oxygen is in Group 6 so O has valency = 2.

Top Tip
Use your Data Booklet to help you find the valencies of different elements.

Using valencies to work out chemical formula

The **chemical formula** of a compound is obtained by swapping over the valencies of the two elements.

To work out the formula of sodium sulphide

Sodium is in Group 1 and so has valency = 1

Sulphur is in Group 6 and so has valency = 2

Top Tip
You must practise writing formulae as this is very important.

We write this as $\overset{1}{Na}\ \overset{2}{S}$ and when the valencies are swapped over we get the formula Na_2S_1. The 1 is not usually written, and so the formula would be correctly written as Na_2S.

If both elements have the same valencies, then the valencies cancel out and the formula contains one atom of each element. An example of this is magnesium oxide, in which both magnesium and oxygen have valency = 2. The formula is written as MgO, not Mg_2O_2.

Formulae of compounds containing group ions

Group ions contain two or more atoms and usually have a negative charge. The formulae of these ions are given on page 4 of the Data Booklet. We can take the number of charges on the ion as the valency of the ion:

- If the ion has a charge of 1+ or 1− then it has valency = 1.
- If the charge is 2− then the valency is 2.
- If the charge is 3−, then the valency is 3.

The chemical formula is obtained by swapping over the valencies as above.

Quick Test

1. Write the formulae for:
   a) sulphur trioxide
   b) carbon tetrachloride.

2. Write down the valencies of the following elements:
   a) chlorine
   b) nitrogen.

3. Write the chemical formula for:
   a) sodium oxide
   b) magnesium sulphide
   c) aluminium oxide.

4. Write the chemical formula for:
   a) sodium carbonate
   b) magnesium sulphate
   c) potassium nitrate.

Answers 1. a) SO_3; b) CCl_4. 2. a) 1; b) 3. 3. a) Na_2O; b) MgS; c) Al_2O_3. 4. a) Na_2CO_3; b) $MgSO_4$; c) KNO_3.

Formulae: using Roman numerals and brackets

Using Roman numerals to show the valency

Some elements, particularly the transition metals, do not always have the same valency in their different compounds. An example is copper, which usually has valency of 2, but in some compounds it has valency of 1. The valency of these elements is usually given in Roman numerals inside brackets. Examples of Roman numerals are given in the table below.

Transition metal valency	Roman numeral
1	(I)
2	(II)
3	(III)
4	(IV)
5	(V)

- In copper(I) oxide the valency of copper is 1, and since oxygen has valency of 2, the formula is Cu_2O.
- The formula of copper(II) oxide is CuO, since both copper and oxygen have the same valency of 2.
- The formula of copper(II) sulphate is $CuSO_4$, since both copper and sulphate have valency of 2.
- Iron usually has valency 2 or 3. For example, iron(II) oxide is FeO and iron(III) oxide is Fe_2O_3.

Note that although both are transition elements, zinc always has valency 2 and silver always has valency 1.

It is not just transition metals that can have different valencies. For example, phosphorus(III) chloride is PCl_3 and phosphorus(V) chloride is PCl_5.

Top Tip

Chemical formulae are important. It is important that you can write them correctly. Using the Data Booklet should help you do this.

Formulae containing brackets

Sometimes brackets are needed in formulae to show that you need more than one ion.

An example is calcium nitrate:

- Calcium is in Group 2 and has valency of 2.
- Nitrate is NO_3 and has valency of 1.
- When the valencies are swapped over there will be one calcium in the formula and two nitrates. The formula must be written with nitrate inside brackets as $Ca(NO_3)_2$ to show that we need to $2 \times NO_3$ to balance the formula. Writing the formula in this way shows that it contains one calcium, two nitrogen and six oxygen atoms. The 2 outside the brackets shows that everything inside the brackets is multiplied by 2.

Other examples include:

- Calcium hydroxide has the formula $Ca(OH)_2$, as calcium has valency 2 and hydroxide has valency 1.
- Ammonium sulphate has the formula $(NH_4)_2SO_4$. Ammonium, NH_4, has valency 1 and sulphate, SO_4, has valency 2. When the valencies are swapped over, the formula for the ammonium ion must be in brackets as it has to be multiplied by 2. The formula for sulphate does not have to be in brackets because ammonium has a valency of 1 and therefore sulphate is multiplied by 1.

Top Tip

When any of the formulae of the ions given in the table at the foot of page 4 of the Data Booklet are multiplied by 2 or more, then the formulae of these ions must be put inside brackets.

Quick Test

1. Write chemical formulae for the following substances:
   a) iron(III) sulphide
   b) cobalt(II) chloride
   c) nickel(II) oxide
   d) lead(II) oxide
   e) lead(IV) oxide
   f) manganese(II) oxide
   g) manganese(IV) oxide
   h) calcium nitrate
   i) lead(II) nitrate
   j) magnesium hydroxide
   k) ammonium phosphate
   l) ammonium carbonate
   m) nickel(II) carbonate
   n) nitrogen trichloride
   o) chromium(III) chloride
   p) chromium(III) oxide
   q) vanadium(V) oxide
   r) titanium(IV) chloride
   s) aluminium hydroxide
   t) barium hydroxide
   u) calcium phosphate
   v) iron(III) sulphide
   w) copper(I) oxide
   x) iron(III) carbonate
   y) magnesium nitrate
   z) iron(III) sulphate

2. Write chemical formulae for the following substances:
   a) oxygen gas
   b) silver(I) nitrate
   c) magnesium fluoride
   d) calcium sulphate
   e) sodium sulphite
   f) aluminium sulphate
   g) carbon monoxide
   h) nitrogen dioxide
   i) phosphorus trichloride
   j) lead(II) carbonate
   k) silicon dioxide
   l) lead(II) bromide
   m) aluminium sulphite
   n) magnesium phosphate
   o) lithium sulphate

Answers 1. a) Fe_2S_3 **b)** $CoCl_2$ **c)** NiO **d)** PbO **e)** PbO_2 **f)** MnO **g)** MnO_2 **h)** $Ca(NO_3)_2$ **i)** $Pb(NO_3)_2$ **j)** $Mg(OH)_2$ **k)** $(NH_4)_3PO_4$ **l)** $(NH_4)_2CO_3$ **m)** $NiCO_3$ **n)** NCl_3 **o)** $CrCl_3$ **p)** Cr_2O_3 **q)** V_2O_5 **r)** $TiCl_4$ **s)** $Al(OH)_3$ **t)** $Ba(OH)_2$ **u)** $Ca_3(PO_4)_2$ **v)** Fe_2S_3 **w)** Cu_2O **x)** $Fe_2(CO_3)_3$ **y)** $Mg(NO_3)_2$ **z)** $Fe_2(SO_4)_3$ **2. a)** O_2 **b)** $AgNO_3$ **c)** MgF_2 **d)** $CaSO_4$ **e)** Na_2SO_3 **f)** $Al_2(SO_4)_3$ **g)** CO **h)** NO_2 **i)** PCl_3 **j)** $PbCO_3$ **k)** SiO_2 **l)** $PbBr_2$ **m)** $Al_2(SO_3)_3$ **n)** $Mg_3(PO_4)_2$ **o)** Li_2SO_4

Balanced equations

Key Facts

- Formula equations show the number of atoms.
- There must be the same number of atoms on both sides of the equation: atoms cannot be created or destroyed.

magnesium + oxygen → magnesium oxide

$$2Mg \quad + \quad O_2 \quad \rightarrow \quad 2MgO$$

State symbols

State symbols can be added to an equation to give **extra information**. They show what state the reactants and products are in. The symbols are:

- (s) for solid
- (l) for liquid
- (g) for gas
- (aq) for aqueous, or dissolved in water

Example

magnesium + oxygen → magnesium oxide

$$2Mg(s) \quad + \quad O_2(g) \quad \rightarrow \quad 2MgO(s)$$

Word equations

What happens in a chemical reaction can be summarised in a word equation. The **reactants** are written on the left hand side of the equation and the **products** are written on the right hand side of the equation. This is shown as:

reactants → products

where the → means 'changes into'.

For example, when magnesium burns in air the magnesium is reacting with oxygen in the air making magnesium oxide. The word equation is:

magnesium + oxygen → magnesium oxide

Formula equations

When **hydrogen** burns in **oxygen**, **water** is made. The word equation is:

hydrogen + oxygen → water

Putting in the correct formulae gives the formula equation:

$$H_2 \quad + \quad O_2 \quad \rightarrow \quad H_2O$$

The **formulae** are correct, but the c=equation is **not** balanced because there are different numbers of atoms on each side of the equation.

The formulae **cannot** be changed, but the numbers in front of the formulae **can** be changed.

Top Tip
Remember that hydrogen and oxygen both exist as diatomic molecules.

How to balance an equation

$$H_2 + O_2 \rightarrow H_2O$$

Looking at the equation for hydrogen burning in oxygen we can see that there are **two** oxygen atoms on the left-hand side but only **one** on the right-hand side.

So a **2** is placed **in front** of the H_2O:

$$H_2 + O_2 \rightarrow 2H_2O$$

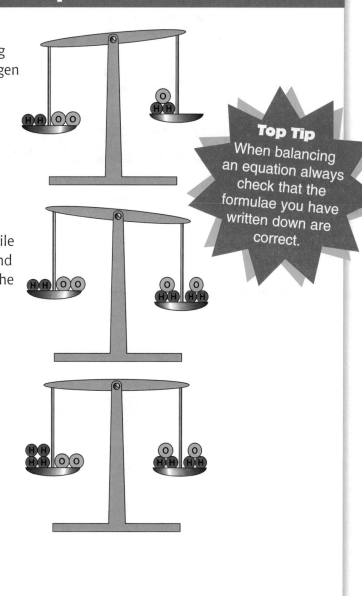

Top Tip
When balancing an equation always check that the formulae you have written down are correct.

Now the oxygen atoms are balanced, but while there are two hydrogen atoms on the left-hand side there are now four hydrogen atoms on the right-hand side.

So a 2 is placed in front of the H_2:

$$2H_2 + O_2 \rightarrow 2H_2O$$

The equation is then balanced.

Top Tip
If you have to write the equation for a reaction, it may be easier to write it in words first.

Quick Test

1. How many calcium atoms are present in $CaCO_3$?

2. How many carbon atoms are present in $CaCO_3$?

3. How many oxygen atoms are present in $CaCO_3$?

4. Why must there be the same number of atoms on both sides of the equation?

5. Balance the equation
$Na(s) + Cl_2(g) \rightarrow NaCl(s)$.

6. Balance the equation
$H_2(g) + Cl_2(g) \rightarrow HCl(g)$.

7. Balance the equation
$C(s) + CO_2(g) \rightarrow CO(g)$.

8. What does the state symbol (l) indicate?

9. What does the state symbol (aq) indicate?

10. Add the state symbols to this equation for the thermal decomposition of calcium carbonate:
$CaCO_3 \rightarrow CaO + CO_2$

Calculations involving number of moles and using balanced equations

Relative formula mass

The **relative formula mass** of any substance is worked out by adding together the **relative atomic masses** of all the atoms in the formula of that substance.

- For water, H_2O:

The formula mass of H_2O is **18**.

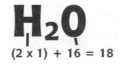

H_2O
$(2 \times 1) + 16 = 18$

Top Tip
Remember that the relative atomic masses of the more common elements are given in the table on page 4 of the Data Booklet.

The mole

No units are given for relative formula masses. Very often in chemistry we need to know actual quantities. To do this we use a value known as **the mole**. The mole is the **gram formula mass** of a substance. This is really the same as the relative formula mass, but with grams as the unit. The word "**mole**" is often abbreviated to "**mol**".

- 1 mole of water, H_2O, is 18 g.
- 0.5 moles of water will be $18 \times 0.5 = 9$ g.

Moles to mass and mass to moles

To do some calculations it is helpful to know and use the triangle below. This shows the relationship between the number of moles, n, the mass and formula mass, FM, of a substance.

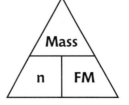

n = number of moles
FM = formula mass

Using this formula
$n = Mass/FM$
$Mass = n \times FM$
$FM = mass/n$

Examples

Question 1:   Calculate the number of moles in 10 g of sodium hydroxide.

Worked answer: Sodium hydroxide has formula NaOH
Formula Mass, FM, of NaOH = 23 + 16 + 1 = 40
Mass = 10 g
n = mass/FM = 10/40 = **0.25 moles**

Question 2: Calculate the mass of 2 moles of calcium chloride.

Worked answer: Calcium chloride has formula $CaCl_2$
Formula mass, FM, of $CaCl_2$ = 40 + (35.5 × 2) = 111
Number of moles, n = 2
Mass = n × FM = 2 × 111 = **222 g**

Calculating the mass of a product

The **balanced formula equation** for a chemical reaction can be used to calculate how much product is made, starting from a known mass of reactant.

Example

What mass of water is produced when 8 g of hydrogen is burned in oxygen?

First you need to work out the balanced formula equation.
$$2H_2(g) + O_2(g) \longrightarrow 2H_2O(g)$$
This tells us that 2 moles of H_2 reacts with 1 mole of O_2 gas to produce 2 moles H_2O.

You then need to work out the relative formula masses and convert them into moles:

- For H_2, this is 1 x 2 = 2, so 2 moles = 4 g
- For O_2, this is 16 x 2 = 32, so 1 mole = 32 g
- For H_2O, this is (2x1) + 16 = 18, so 2 moles = 36 g

So 4 g of H_2 produces 36 g of H_2O and therefore 1 g of H_2 produces 36/4 g of H_2O = 9 g.

Therefore 8 g of H_2 produces 9 x 8 = 72 g.

This calculation shows that if 8 g of H_2 is burned in O_2 then 72 g of H_2O is made.

Top Tip
When you calculate the masses of the reactants and products you will see that overall no mass has been gained or lost.

Quick Test

1. Calculate the mass of:
   a) 2 moles of $Ca(OH)_2$ b) 0.3 moles of $CuSO_4$

2. Calculate the number of moles in:
   a) 1.0 g of $CaCO_3$ b) 13.2 g of CO_2

3. Calculate the mass of water produced when 0.2 g of hydrogen is burned in air.

4. The equation for calcium burning in oxygen is $2Ca(s) + O_2(g) \longrightarrow 2CaO(s)$.
   Calculate the mass of calcium required to produce 2.24 g of calcium oxide.

Answers: 1. a) 148 g **b)** 47.85 g. **2. a)** 0.01 mol **b)** 0.3 mol. **3.** 1.8 g. **4.** 1.6 g.

PPA 1: The effect of concentration changes on reaction rate

Introduction

When sodium persulphate solution is added to potassium iodide solution, iodine is formed. If small quantities of starch and sodium thiosulphate are included, the reaction mixture is initially colourless, but after some time, it suddenly turns blue/black as the starch reacts with the iodine. The time taken for the blue/black colour to appear can be taken as the reaction time, t, and the rate will be 1/t .

Aim

To find the effect of varying the concentration of sodium persulphate solution on its rate of reaction with potassium iodide solution.

Procedure

10 cm^3 of sodium persulphate solution and 1 cm^3 of starch solution were measured into a dry glass beaker placed on a white tile. 10 cm^3 of potassium iodide solution (containing sodium thiosulphate) was quickly added and the timer started. The time for the mixture to suddenly go blue/black was noted. The procedure was repeated three times with different concentrations of sodium persulphate. The concentration of the sodium persulphate solution was reduced by taking less than 10 cm^3 of it and making up the difference with water.

Results

Volume of sodium persulphate solution (cm^3)	Volume of water (cm^3)	Reaction time (s)	Reaction rate (s^{-1})
10	0	42	0·0238
8	2	53	0·0189
6	4	71	0·0141
4	6	105	0·0095

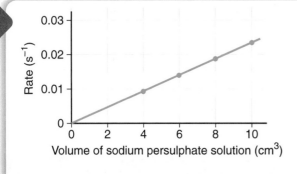

Conclusion

As the concentration of the sodium persulphate solution increases, the reaction rate increases.

Points to note

- Only the concentration of the sodium persulphate solution was varied. Other factors, like temperature and the concentrations and volumes of the other reactants had to be kept constant.
- The amount of iodine formed before the reaction mixture turned blue/black was the same in each experiment.
- Since the colour change to blue/black was very sharp, the reaction time measurement was accurate.
- A white tile was used to make it easier to see the colour change.
- The total volume of the reaction mixture was the same in each experiment. This ensured that the concentrations of the other reactants were kept constant. It also meant the volume of sodium persulphate solution could be taken as a measure of its concentration.

Quick Test

1. What is being changed in this experiment?
2. In the experiment the stopwatch is started when the reactants are mixed. When should the stopwatch be stopped?
3. Write down three factors that were kept the same in the experiment.
4. What do the results of the experiment indicate?
5. What was used to help make the colour change easier to see?

Answers 1. The concentration of the sodium persulphate solution. **2.** When the blue/black colour appears. **3.** The total volume of the solution, the temperature, the concentrations and volumes of the other reactants. **4.** As the concentration of the sodium persulphate increases, the reaction rate increases. **5.** Placing the beaker on a white tile.

PPA 2: The effect of temperature changes on reaction rate

Introduction

When an acid is added to sodium thiosulphate solution, sulphur is formed. Initially the reaction mixture is clear, but gradually it becomes cloudy as solid sulphur appears. The time taken for the sulphur to obscure a cross on a piece of paper can be taken as the reaction time, t, and the rate will be 1/t.

Aim

To find the effect of varying temperature on the rate of reaction between sodium thiosulphate solution and hydrochloric acid.

Procedure

20 cm³ of sodium thiosulphate solution was measured into a dry glass beaker and placed on a piece of paper with a cross marked on it. 1 cm³ of hydrochloric acid was added to the sodium thiosulphate solution and the timer started. The time taken for the cross to be obscured was noted and the temperature of the mixture was taken. The experiment was repeated at 30°C, 40°C and 50°C, with the temperature of the reaction mixture being accurately noted at the completion of each experiment.

Results

Temperature (°C)	Reaction time (s)	Rate (s⁻¹)
19	71	0·0141
28	45	0·0222
39	25	0·0400
49	13	0·0769

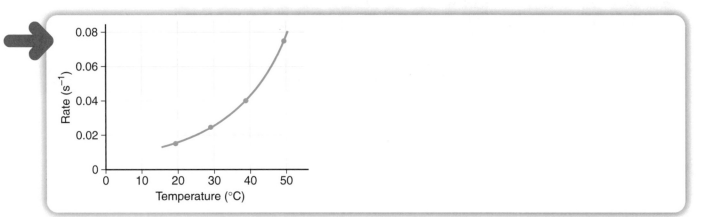

Conclusion

As the temperature increases the reaction rate increases.

Points to note

- Only the temperature of the reaction mixture was varied. Other factors, like the concentrations and volumes of the reactants had to be kept constant.
- The same or identical beakers were used to carry out the experiments so that the depth of the solution was kept constant.
- In each experiment, the same cross was used and the same person viewed the cross, since eyesight varies from person to person.
- The amount of sulphur needed to obscure the cross was the same in each experiment, but the time taken for this to happen varied.
- Sulphur dioxide is a product of the reaction, and since it irritates the lungs, the experiments were carried out in a well ventilated room and the sodium thiosulphate solution was not heated beyond 50°C.

Quick Test

1. What was the aim of the experiment?
2. What was measured during the reaction?
3. When was the temperature measured accurately?
4. What was the reaction rate taken as?
5. Why should this experiment be carried out in a well ventilated room?
6. What happens to the reaction rate as the temperature increases?

Answers 1. To find the effect of changing the temperature on the rate of the reaction between sodium thiosulphate solution and hydrochloric acid. **2.** The time taken for the cross to be obscured by the sulphur produced in the reaction. **3.** At the completion of each experiment. **4.** 1/the time taken for the cross to be obscured. **5.** Because the sulphur dioxide produced will irritate the lungs. **6.** The reaction rate also increases.

PPA 3: Electrolysis

Introduction

Passing a current of electricity through a solution of an ionic compound, i.e. an electrolyte, is known as electrolysis. The ions present in the electrolyte undergo chemical changes at the electrodes, and this may lead to the break up of the ionic compound.

Aim

To electrolyse copper(II) chloride solution and to identify the products at the electrodes.

Procedure

Using the apparatus shown below, a D.C. current of electricity was passed through copper(II) chloride solution for a few minutes.

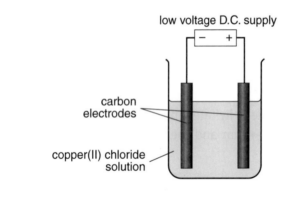

Results

Electrode	Observations
negative	brown solid formed
positive	bubbles of gas formed; pungent smell; blue litmus paper bleached

Conclusion

Copper(II) chloride solution was electrolysed, producing copper at the negative electrode and chlorine at the positive electrode.

Points to note

- Direct current (D.C.) had to be used so that only one product was formed at each electrode.
- Carbon electrodes were used because carbon (graphite) is a good conductor of electricity and, since it is unreactive, it will not react with the copper(II) chloride solution.
- Since chlorine is toxic, it had to be smelled very cautiously. With the nose at least 30 cm from the cell, the gas was wafted in the direction of the nose and just a sniff was taken.

Quick Test

1. What is meant by the term electrolysis?

2. What is meant by the term electrolyte?

3. During the electrolysis of copper(II) chloride solution, what is observed at;
   a) the negative electrode
   b) the positive electrode?

4. What happened to blue litmus paper held at the positive electrode?

5. Which electrodes were used? Write down two reasons why these electrodes were used.

6. Indicate how the gas produced should be smelled.

Combustion

What is a fuel?

A **fuel** is a **chemical** which is burned to produce **energy**. Examples of fuels include **coal**, **oil** and **natural gas** known as the **fossil fuels**. Burning fossil fuels satisfies most of the world's energy needs, but they are **finite resources** and will eventually run out.

Burning is a **chemical reaction** in which a substance reacts with the oxygen in the air. The reaction is also known as **combustion** and in all combustion reactions, energy is released.

Hydrocarbons

The main compounds found in oil and natural gas are **hydrocarbons**. A hydrocarbon is a compound that only contains hydrogen and carbon.

Top Tip
Remember that hydrocarbons contain only hydrogen and carbon.

Complete combustion of hydrocarbons

Top Tip
Remember the test for carbon dioxide – it turns limewater cloudy.

When hydrocarbons burn in a plentiful supply of air, they undergo **complete combustion** to form water and carbon dioxide. Under these conditions there is enough oxygen present for all of the hydrogen to be converted into water (hydrogen oxide) and all of the carbon to be converted into carbon dioxide. The following equation shows the complete combustion of methane (CH_4), the main constituent of natural gas.

$$CH_4 + 2O_2 \rightarrow CO_2 + 2H_2O$$

The following apparatus can be used to demonstrate that the combustion of a hydrocarbon produces water and carbon dioxide.

The liquid that collects in the first test tube freezes at 0°C and boils at 100°C, so must be water. The gas that bubbles through the second test tube turns the **limewater** cloudy and so must be carbon dioxide since this is the test for carbon dioxide. In general terms, when a substance is burned and water (hydrogen oxide) is produced, this proves the substance must contain hydrogen. Similarly, if carbon dioxide is produced, the substance must contain carbon.

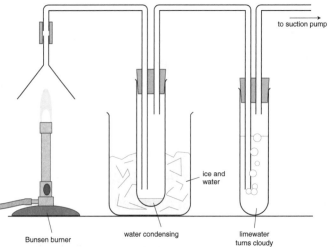

to suction pump

Bunsen burner

water condensing

ice and water

limewater turns cloudy

Incomplete combustion of hydrocarbons

When hydrocarbons burn in a limited supply of air, they undergo **incomplete combustion** to form carbon (soot), carbon monoxide and water. The hydrogen is still converted into water, but there is not enough oxygen to convert the carbon into carbon dioxide. Instead, carbon (soot) and the very poisonous gas, carbon monoxide, are produced.

Air pollution

A **pollutant** is any substance that damages the environment. Most pollutants in the air come from burning fossil fuels.

Carbon monoxide

When hydrocarbons in fuels undergo incomplete combustion, one of the products is the poisonous gas **carbon monoxide** (CO).

Oxides of nitrogen

In the cylinders of petrol engines, the petrol/air mixture is sparked. As well as igniting the petrol, the spark provides enough energy to allow some of the nitrogen and oxygen to react to form oxides of nitrogen such as **nitrogen dioxide** (NO_2) and **nitrogen monoxide** (NO). These gases are poisonous and nitrogen dioxide is a contributor to acid rain.

Sulphur dioxide

Many fuels contain small amounts of sulphur-containing compounds, and when they burn, the poisonous gas, **sulphur dioxide** (SO_2) is produced. Sulphur dioxide is the main contributor to acid rain.

Soot (carbon) particles

When hydrocarbons undergo incomplete combustion, particles of soot (carbon) are produced. This type of pollution is particularly pronounced when **diesel** fuel is burned.

One way of reducing air pollution caused by the combustion of hydrocarbons in cars is to fit **catalytic converters** to their exhausts. The **transition metal catalysts** present in the converters speed up the conversion of the pollutant gases into less harmful gases – carbon monoxide into carbon dioxide, nitrogen oxides into nitrogen and oxygen, and unburned hydrocarbons into water and carbon dioxide.

Quick Test

1. What is a fuel?
2. Name the products of the complete combustion of a hydrocarbon.
3. What is the test for carbon dioxide?
4. Name the products of the incomplete combustion of a hydrocarbon.
5. What does a catalytic converter do?

Answers 1. A chemical that is burned to produce energy. **2.** Water and carbon dioxide. **3.** It turns limewater cloudy. **4.** Water, carbon monoxide and carbon. **5.** It converts pollutant gases into less harmful gases.

Fractional distillation

Crude oil

Crude oil is a mixture of hundreds of different compounds, the vast majority of which are **hydrocarbons**. Unlike the other fossil fuels, crude oil can not be used directly as a fuel. It has to be separated into so-called **fractions**, before it is of any use.

Fractional distillation of crude oil

Crude oil can be separated into fractions by a process called **fractional distillation**. It is the differences in the boiling points of the hydrocarbons that allow the fractions to be separated. A **fraction** is a group of hydrocarbons with boiling points within a given range. The information below shows a possible result from fractional distillation.

Fraction	Temperature range (°C)	Number of carbon atoms per molecule
Refinery gas	<20	1–4
Naphtha	20–130	5–12
Kerosene	130–240	11–16
Gas oil	240–350	15–25
Residue	>350	>25

fractionating tower

heated crude oil

The crude oil is heated before it enters the fractionating tower and at this stage most of the hydrocarbons are in the gaseous state. As each hydrocarbon gas rises up the tower, it reaches a point where the temperature in the tower is just below its boiling point and it **condenses** into a liquid. Since the hydrocarbons in crude oil have different boiling points they will condense at different points within the tower. As a result, they are separated into fractions. Each fraction still contains a mixture of hydrocarbons, but their boiling points and molecular sizes cover a narrower range.

Uses of the fractions

Fraction	Uses
Refinery gas	Fuels for cooking and heating
Naphtha	To make petrol and other chemicals
Kerosene	Aviation and heating fuels
Gas oil	To make diesel
Residue	To make fuel oil for ships and power stations; lubricating oils and waxes; bitumen for roads and roofing

Properties of the fractions

The following table summarises the trends in some of the **properties** of the fractions separated from crude oil.

Fraction	Molecular size	Boiling point	Ease of evaporation	Viscosity	Flammability
Refinery gas	Small	Low	Easy	Low	High
Naphtha					
Kerosene					
Gas oil					
Residue	Large	High	Difficult	High	Low

As we move from the refinery gas down to the residue the size of the hydrocarbon molecules increases and it is these differences in **molecular size** that allow us to explain the trends in the properties of the fractions.

Boiling point

When a liquid hydrocarbon boils, the forces of attraction between its molecules have to be broken. As the molecules get bigger, there are more of these forces and so the boiling point increases.

Ease of evaporation

The ease with which a liquid hydrocarbon evaporates also depends on the forces of attraction between its molecules. The smaller the molecules, the fewer the forces between them and the easier it is for the hydrocarbon to evaporate.

Viscosity

The viscosity of a liquid is a measure of how 'thick' it is. The bigger the molecules, the more they tangle up with each other, and the more viscous is the liquid.

Flammability

A flammable hydrocarbon is one that catches fire easily. It will have a low boiling point and evaporate easily. The smaller the molecules, the more flammable it is.

The uses of the fractions are related to the above properties.

Top Tip
The properties of the fractions separated from crude oil can be explained in terms of molecular size.

Quick Test

1. What is a fraction?
2. Why can the hydrocarbons in crude oil be separated into fractions?
3. In which fraction will octane (C_8H_{18}) be found?
4. Why do the hydrocarbons in kerosene have higher boiling points than those in refinery gas?
5. Which fraction is more viscous, residue or naphtha?

Answers 1. A group of hydrocarbons with boiling points within a given range. **2.** Because they have different boiling points. **3.** Naphtha. **4.** Because the molecules are bigger and there are more forces of attraction between the molecules. **5.** Residue.

Hydrocarbons

Alkanes

There are an enormous number of different hydrocarbons but they can be divided into just a few subsets. The family of **alkanes** is one subset of hydrocarbons. The following table shows the names and molecular formulae of the first eight members of the alkane family.

Name of alkane	methane	ethane	propane	butane	pentane	hexane	heptane	octane
Molecular formula	CH_4	C_2H_6	C_3H_8	C_4H_{10}	C_5H_{12}	C_6H_{14}	C_7H_{16}	C_8H_{18}

The names of these hydrocarbons all end in **-ane** and this identifies them as belonging to the alkane family. The **molecular formula** gives the number of carbon and hydrogen atoms in each alkane molecule. The pattern in the molecular formulae shows that the general formula for the alkanes is C_nH_{2n+2} where **n** is the number of carbon atoms.

Two types of formulae can be used to show the structures of alkanes; a **full structural formula** and a **shortened structural formula**. These are drawn below for three members of the alkane family.

Alkane	Molecular formula	Full structural formula	Shortened structural formula
Methane	CH_4	H H–C–H H	CH_4
Pentane	C_5H_{12}	H H H H H H–C–C–C–C–C–H H H H H H	$CH_3CH_2CH_2CH_2CH_3$ or $CH_3–CH_2–CH_2–CH_2–CH_3$
Octane	C_8H_{18}	H H H H H H H H H–C–C–C–C–C–C–C–C–H H H H H H H H H	$CH_3CH_2CH_2CH_2CH_2CH_2CH_2CH_3$ or $CH_3–CH_2–CH_2–CH_2–CH_2–CH_2–CH_2–CH_3$

The full structural formula shows every atom and every bond in the molecule. The shortened structural formula gives the same information as the full structural formula, but shows no bonds or just the carbon-to-carbon bonds.

All alkanes contain carbon-to-carbon single bonds and those we have considered so far are known as **straight-chain alkanes** because the carbon atoms are joined in one continuous chain.

Top Tip

The names of the first eight alkanes are given on page 6 of your Data Booklet. You should be able to write their molecular formulae and draw their structural formulae.

The full structural formula and shortened structural formulae for a typical **branched-chain alkane** are drawn opposite.

The molecular formula for this alkane is C_8H_{18} but we cannot call it octane because this is the name for the straight-chain alkane. We can regard it as being a straight-chain of five C atoms with three $-CH_3$ branches (shaded) attached. The $-CH_3$ branch is just methane (CH_4) minus a H atom and it is called a **methyl branch**. Similarly, a $-CH_2CH_3$ branch would be called **ethyl**. To work out the name, we apply the following rules.

or

$$CH_3-CH-C-CH_2-CH_3$$

or

$$CH_3CH(CH_3)C(CH_3)_2CH_2CH_3$$

- Pick out the longest carbon chain to get the parent name. In our example, the parent name must be **pentane** because it contains 5 C atoms.

- Number the C atoms in this chain starting at the end nearer a branch. Numbering therefore starts at the left-hand end in this example.

- Identify the branches and arrange them in alphabetical order. For example, ethyl will come before methyl. If there are two or more of the same branch, indicate this by using di, tri and so on. Then identify the numbers of the C atoms in the chain at which the branches are attached. In our example:
  — there is only one type of branch   **methyl**
  — but there are three of them   **trimethyl**
  — and they are attached at C-2, C-3 and C-3   **2,3,3-trimethyl**

- We therefore write 2,3,3-trimethyl in front of the parent name and arrive at **2,3,3-trimethylpentane** as the systematic name.

> **Top Tip**
> Make sure you learn the rules for naming branched-chain alkanes.

As well as working out the name of a branched-chain alkane, you need to be able to draw a structural formula for an alkane given its name. Take **2,4-dimethylhexane**, for example. The parent name is **hexane**, so we draw a chain of 6 C atoms and then attach two methyl branches, one at C-2 and another at C-4. Finally, we add H atoms and get the structure drawn opposite:

Quick Test

1. Alkanes are a subset of which larger set of compounds?

2. How many carbon atoms would be present in an alkane containing 20 hydrogen atoms?

3. Name the straight chain alkane with molecular formula C_4H_{10} and draw its full structural formula.

4. Name the following branched-chain alkane: CH_2CH_3
$$CH_3 - CH_2 - CH - CH_2 - CH_3$$

5. Draw the full structural formula for 2,2,3-trimethylbutane.

Answers 1. Hydrocarbons. **2.** Nine. **3.** Butane. **4.** 3-ethylpentane. **5.**

More hydrocarbons

Like the alkanes, the **alkenes** are also a subset of hydrocarbons. An alkene, however, differs from an alkane because it contains a carbon-to-carbon double bond within its structure. The following table shows the names, molecular formulae and full and shortened structural formulae for the first two members of the alkene family.

Alkene	Molecular formula	Full structural formula	Shortened structural formula
Ethene	C_2H_4	H–C=C–H H H	$CH_2=CH_2$
Propene	C_3H_6	H–C=C–C–H (with H atoms)	$CH_2=CHCH_3$ or $CH_2=CH–CH_3$

The names of these hydrocarbons all end in **-ene** and this identifies them as belonging to the alkene family. The presence of a carbon-to-carbon double bond in their structural formulae also identifies them as alkenes. The pattern in their **molecular formulae** implies that the general formula for the alkenes is C_nH_{2n}.

Naming straight-chain alkenes

There are **three straight-chain alkenes** with molecular formula, C_6H_{12}. Their full structural formulae are drawn below and labelled **X**, **Y** and **Z**.

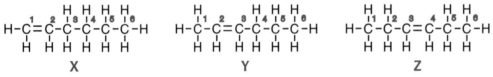

$$\underset{X}{H-\overset{1}{C}=\overset{2}{C}-\overset{3}{C}-\overset{4}{C}-\overset{5}{C}-\overset{6}{C}-H} \qquad \underset{Y}{H-\overset{1}{C}-\overset{2}{C}=\overset{3}{C}-\overset{4}{C}-\overset{5}{C}-\overset{6}{C}-H} \qquad \underset{Z}{H-\overset{1}{C}-\overset{2}{C}-\overset{3}{C}=\overset{4}{C}-\overset{5}{C}-\overset{6}{C}-H}$$

Alkenes **X**, **Y** and **Z** differ in the positions of the carbon-to-carbon double bond. To name them, we apply the following rules:

- Pick out the longest carbon chain containing the double bond to get the parent name. For **X**, **Y** and **Z** the longest chain containing the double bond has 6 C atoms and so the parent name for all three is **hexene**.
- Number the C atoms in this chain starting at the end nearer the double bond.
- Identify the number of the C atom where the double bond starts and insert it into the name:
    — For **X** the double bond starts at C-1 and so it is called **hex-1-ene**
    — For **Y** the double bond starts at C-2 and so it is called **hex-2-ene**
    — For **Z** the double bond starts at C-3 and so it is called **hex-3-ene.**

As well as naming alkenes, you need to be able to draw the structural formula for a straight-chain alkene given its name. Take **pent-2-ene**, for example. The parent name is **pentene**, so we draw a chain of 5 C atoms and then insert a double bond starting at C-2. Numbering can start from either end just as long as the double bond starts at C-2. Finally, we add H atoms and get:

$$H-\overset{H}{\underset{H}{C}}-C=C-\overset{H}{\underset{H}{C}}-\overset{H}{\underset{H}{C}}-H$$

Cycloalkanes

The **cycloalkanes** are another subset of hydrocarbons. As the name implies, the carbon atoms in a cycloalkane are joined in a ring rather than a chain. To form a ring, there has to be a minimum of three carbon atoms and this means that cyclopropane must be the simplest cycloalkane. The names, molecular formulae and structural formulae of two members of the cycloalkane family are shown in the following table.

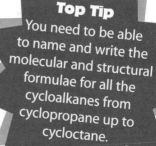

Top Tip
You need to be able to name and write the molecular and structural formulae for all the cycloalkanes from cyclopropane up to cyclooctane.

Cycloalkane	Molecular formula	Full structural formula	Shortened structural formula
Cyclopropane	C_3H_6		
Cyclopentane	C_5H_{10}		

The names of these hydrocarbons all start with **cyclo-** and end with **-ane** which identifies them as belonging to the cycloalkane family. The cycloalkanes have the general formula, C_nH_{2n} which is the same as that for the alkenes. The carbon-to-carbon bonds in the cycloalkanes are all single covalent bonds just as they are in the alkanes.

Homologous series

A homologous series is a family of compounds whose members have the same general formula and similar chemical properties, i.e. they take part in the same type of chemical reactions. The alkanes, alkenes and cycloalkanes are all examples of different homologous series.

Quick Test

1. How many hydrogen atoms would be present in an alkene containing ten carbon atoms?
2. Name the following alkene: $CH_3 - CH = CH - CH_3$
3. Draw a shortened structural formula for oct-3-ene.
4. How many carbon atoms would be present in a cycloalkane containing ten hydrogen atoms?
5. Name the cycloalkane with molecular formula, C_4H_8.

Answers 1. 20. **2.** But-2-ene. **3.** $CH_3CH_2CH = CHCH_2CH_2CH_2CH_3$ or $CH_3-CH_2-CH = CH_2-CH_2-CH_2CH_2-CH_3$ **4.** Five. **5.** Cyclobutane.

Isomers

What are isomers?

Isomers are compounds with the same molecular formula but different structural formulae.

Consider the following compounds:

$$H - \overset{\overset{\displaystyle H}{|}}{\underset{\underset{\displaystyle H}{|}}{C}} - \overset{\overset{\displaystyle H}{|}}{\underset{\underset{\displaystyle H}{|}}{C}} - O - H \qquad H - \overset{\overset{\displaystyle H}{|}}{\underset{\underset{\displaystyle H}{|}}{C}} - O - \overset{\overset{\displaystyle H}{|}}{\underset{\underset{\displaystyle H}{|}}{C}} - H$$

They have different structures because their atoms are bonded together in different ways, but they have the same molecular formula, C_2H_6O. These two compounds are therefore isomers.

Since isomers have different structures they will have different physical properties. Isomers will also have different chemical properties if they belong to different homologous series like the two above. Their chemical properties, however, will be similar if they belong to the same homologous series.

Alkane isomers

There are no isomers of methane (CH_4), ethane (C_2H_6) or propane (C_3H_8) because only one structure can be drawn for each of these molecular formulae. There are, however, two alkane isomers with the molecular formula C_4H_{10} and their full structural formulae and names are given on the right.

butane 2-methylpropane

In butane, the four carbon atoms are all in one long chain, but in 2-methylpropane, there are three carbon atoms in the longest chain with a methyl (-CH_3) branch on the second carbon atom. These compounds have different physical properties: the boiling point of butane is –1°C while that of 2-methylpropane is –12°C. They react in a similar way because they both belong to the same homologous series.

You must be careful when identifying isomers from their structural formulae. Consider, for example, the structural formula on the right.

Although this alkane has the same molecular formula (C_4H_{10}) as butane and 2-methylpropane, it is not an isomer. It is in fact identical to butane because all four carbon atoms are in one long chain. The methyl group (-CH_3) on the left-hand side is not a branch but part of the chain.

As the number of carbon atoms in an alkane increases, the number of isomers increases exponentially. There are over a quarter of a million isomers with the molecular formula $C_{20}H_{42}$!

Alkene and cycloalkane isomers

There are no isomers of the alkenes ethene (C_2H_4) or propene (C_3H_6). There are, however, three alkenes with the molecular formula C_4H_8.

Their structures are shown below.

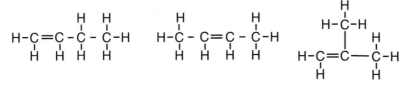

The first two isomers are straight-chain alkenes and the third is a branched-chain alkene. You do not need to know how to name branched-chain alkenes, but you do need to be aware that they are possible isomers.

Since the general formula for alkenes and cycloalkanes is the same (C_nH_{2n}), this means that there must be cycloalkane isomers with molecular formula C_4H_8. Their structures are shown below.

 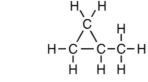

Top Tip

Remember that it is possible for isomers to belong to different homologous series.

The isomer on the right is a branched cycloalkane.

In all, there are five isomers with molecular formula C_4H_8: three of them are alkenes and two are cycloalkanes.

Quick Test

1. What are isomers?

2. For each of the following pairs of compounds shown opposite, state whether they are isomers or identical or neither.

3. Draw full structural formulae for the two isomers with the molecular formula C_3H_6.

4. Name a straight-chain alkene which is an isomer of cyclopentane.

5. Name a cycloalkane isomer of oct-3-ene.

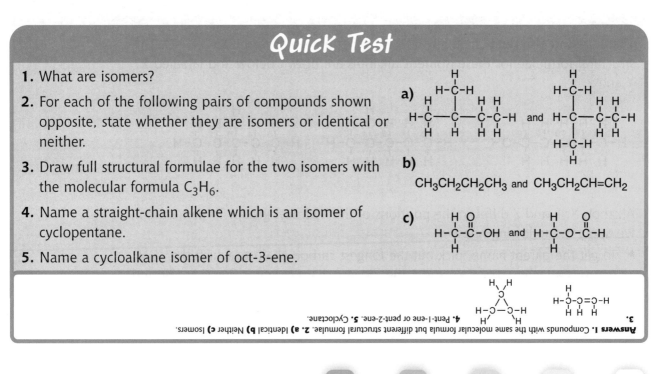

Alkanols and alkanoic acids

Alkanols

The **alkanols** are another family of organic compounds, but unlike the alkane, alkene and cycloalkane families, they are not hydrocarbons, since they contain oxygen in addition to carbon and hydrogen. The oxygen in an alkanol is present in an **–OH** group which is known as the **hydroxyl group**. The names, molecular formulae and structural formulae of the first two members of the alkanol family are shown in the following table.

Alkanol	Molecular formula	Full structural formula	Shortened structural formula
Methanol	CH_4O	H | H–C–O–H | H	CH_3OH or $CH_3–OH$
Ethanol	C_2H_6O	H H | | H–C–C–O–H | | H H	CH_3CH_2OH or $CH_3–CH_2–OH$

Top Tip

Some alkanols are given on page 6 of your Data Booklet. You need to be able to name and write molecular formulae and structural formulae for all the alkanols from methanol up to octanol.

The names of these compounds end in **-ol** and this identifies them as belonging to the alkanol family. The presence of a hydroxyl group (–OH) in their structural formulae also identifies them as alkanols.

Alkanols are a subset of the larger set of compounds called alcohols, and, like the alkanols, all alcohols contain a hydroxyl group (–OH).

Naming straight-chain alkanols

There are three straight-chain alkanols with the molecular formula, $C_5H_{12}O$. The structural formulae for these isomeric alkanols are drawn below and labelled **X**, **Y** and **Z**.

Alkanols X, Y and Z differ in the positions of the hydroxyl group. To name them, we apply the following rules:

- To get the parent name, pick out the longest carbon chain to which the –OH group is attached. For **X**, **Y** and **Z** the longest chain contains five C atoms and so the parent name for all three is **pentanol**.
- Number the C atoms in this chain starting at the end nearer the –OH group.

- Identify the number of the C atom to which the –OH group is attach
  insert it into the name:

— For **X** the –OH group is attached at C-1 and so it is called **pentan-1-ol**

— For **Y** the –OH group is attached at C-2 and so it is called **pentan-2-ol**

— For **Z** the –OH group is attached at C-3 and so it is called **pentan-3-ol**.

The **alkanoic acids** are another family of oxygen-containing organic compounds, but this time the oxygen is contained in a **–COOH group** known as the **carboxyl group**.

The names, molecular formulae and structural formulae of the first two members of the alkanoic acid family are shown in the following table.

Alkanoic acid	Molecular formula	Full structural formula	Shortened structural formula
Methanoic acid	CH_2O_2	O ‖ H–C–O–H	HCOOH
Ethanoic acid	$C_2H_4O_2$	H O ‖ H–C–C–O–H H	CH_3COOH or CH_3–COOH

The names of these compounds end with **-oic acid** and this identifies them as belonging to the alkanoic acid family. The presence of the carboxyl group (–COOH) in their structural formulae also identifies them as alkanoic acids.

The full structural formula for the carboxyl group is often abbreviated to:
O ‖ – C – OH

and so the structural formulae for pentanoic acid and heptanoic acid, for example, can be drawn as:

H H H H O H–C–C–C–C–C–OH H H H H
pentanoic acid

H H H H H H O H–C–C–C–C–C–C–C–OH H H H H H H
heptanoic acid

Alkanoic acids are a subset of the larger set of compounds called carboxylic acids and like the alkanoic acids, all carboxylic acids contain a carboxyl group (–COOH).

Quick Test

1. What name is given to the –OH group in alkanols?

2. Name the following alkanol:
   H H O H H–C–C–C–H H H H

3. Write the molecular formula and draw the full structural formula for heptan-3-ol.

4. What name is given to the –COOH group in alkanoic acids?

5. Name the following alkanoic acid:
   $CH_3CH_2CH_2CH_2CH_2CH_2CH_2COOH$.

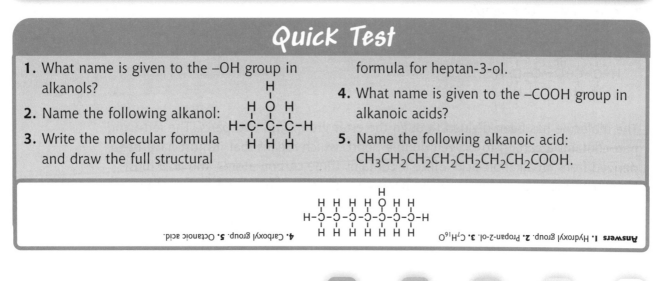

Answers 1. Hydroxyl group. **2.** Propan-2-ol. **3.** $C_7H_{16}O$

H H H H H O H H H–C–C–C–C–C–C–C–H H H H H H H H

4. Carboxyl group. **5.** Octanoic acid.

...ied when an alkanol reacts with an
...te, which is a typical ester, is formed
... The following equations can be used

$$\rightleftharpoons \quad H-\underset{\underset{H}{|}}{\overset{\overset{H}{|}}{C}}-O-\underset{}{\overset{\overset{O}{\parallel}}{C}}-\underset{\underset{H}{|}}{\overset{\overset{H}{|}}{C}}-H + H_2O$$

$$\rightleftharpoons \quad CH_3OOCCH_3 \quad + \quad H_2O$$

acid $\rightleftharpoons$ methyl ethanoate + water

Top Tip
An ester can be identified by the presence of an ester group in its structural formula and the -oate ending in its name.

...e oxygen and hydrogen atoms that form the water molecule have been shaded. You will notice from its structural formula that the ester contains the following group of atoms:

$$-O-\overset{\overset{O}{\parallel}}{C}- \quad \text{or} \quad -OOC-$$

This group is known as the **ester group** or **ester linkage** and its presence identifies a compound as an ester. The **-oate** ending in the name also identifies a compound as an ester.

Naming esters

The name of an ester is related to the names of its parent alkanol and alkanoic acid. The first part of the ester's name comes from the name of the alkanol and the second part from the name of the alkanoic acid. So the ester made from ethanol and propanoic acid will be called ethyl propanoate and that made from methanoic acid and propan-1-ol will be called propyl methanoate.

You must also be able to name an ester given its structural formula. Take the following ester, for example:

$$H-\underset{\underset{H}{|}}{\overset{\overset{H}{|}}{C}}-\underset{\underset{H}{|}}{\overset{\overset{H}{|}}{C}}-\underset{}{\overset{\overset{O}{\parallel}}{C}}-O-\underset{\underset{H}{|}}{\overset{\overset{H}{|}}{C}}-H$$

The molecule has been divided through the ester group into two parts. The left-hand part contains a carbon-to-oxygen double bond, which means that it must have been derived from an alkanoic acid. Since it contains three carbon atoms, the acid must have been propanoic acid. The right-hand part, therefore, has been derived from an alkanol, and since there is only one carbon atom present, the alcohol must have been methanol. So the name of the above ester is methyl propanoate.

The structural formula for an ester can be drawn given the name of the ester. Take butyl methanoate, for example. The best way of working out its structural formula is to write the equation for the reaction between its parent alkanol (butan-1-ol) and its parent alkanoic acid (methanoic acid):

$$
\underset{\text{butan-1-ol}}{H-\overset{\overset{\displaystyle H}{|}}{\underset{\underset{\displaystyle H}{|}}{C}}-\overset{\overset{\displaystyle H}{|}}{\underset{\underset{\displaystyle H}{|}}{C}}-\overset{\overset{\displaystyle H}{|}}{\underset{\underset{\displaystyle H}{|}}{C}}-\overset{\overset{\displaystyle H}{|}}{\underset{\underset{\displaystyle H}{|}}{C}}-O-H} \;+\; \underset{\text{methanoic acid}}{H-O-\overset{\overset{\displaystyle O}{\|}}{C}-H} \;\rightleftharpoons\; \underset{\text{butyl methanoate}}{H-C-C-C-C-O-C-H} \;+\; \underset{\text{water}}{H_2O}
$$

You must be able to draw the structural formula for an ester, given the names of its parent alkanol and alkanoic acid. To do this, you write the equation for the reaction between the two in the same way as that outlined above.

Breaking down esters

When an ester reacts with water it breaks down into its parent alkanol and alkanoic acid and, given its name, you must be able to name the products of the breakdown of an ester. We know that the first part of an ester's name is derived from the parent alkanol and the second part from the parent alkanoic acid. So ethyl pentanoate, for example, will break down into ethanol and pentanoic acid.

You must be able to identify the products of the breakdown of an ester given its structural formula. Consider the ester opposite:

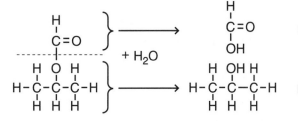

The molecule has been divided through the ester group into two parts. The upper part contains one carbon atom and a carbon-to-oxygen double bond, so will form methanoic acid. The lower part will form propan-2-ol.

Quick Test

1. Name the ester formed from CH_3CH_2COOH and CH_3CH_2OH.

2. Name the following ester:

$$CH_3 - CH_2 - CH_2 - \overset{\overset{\displaystyle O}{\|}}{C} - O - CH_3$$

3. Draw the full structural formula for the ester made from methanol and methanoic acid.

4. Draw the full structural formula for butyl pentanoate.

5. Name the products of the breakdown of octyl ethanoate.

Addition reactions

Saturated and unsaturated compounds

Compounds that contain at least one carbon-to-carbon double bond (C=C) are described as **unsaturated** and those in which all the carbon-to-carbon bonds are single (C-C) are **saturated**. Let's consider a typical member of some of the families of organic compounds we have studied so far and decide whether they are saturated or unsaturated.

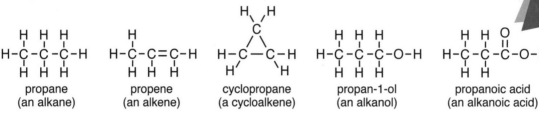

| propane | propene | cyclopropane | propan-1-ol | propanoic acid |
| (an alkane) | (an alkene) | (a cycloalkene) | (an alkanol) | (an alkanoic acid) |

Top Tip
Remember that an unsaturated compound contains at least one C=C double bond.

Only propene is unsaturated because it is the only one that contains a C=C double bond. Propene, however, is not the only alkene which is unsaturated; all alkenes are unsaturated because they all contain a C=C double bond. Alkanes, cycloalkanes, alkanols and alkanoic acids, on the other hand, are saturated.

Testing for unsaturation

It is possible to distinguish an unsaturated compound from a saturated one by using **bromine solution**. Bromine solution is orange-brown in colour, and when an unsaturated compound is added to it, the orange-brown colour rapidly turns colourless. When a saturated compound is added, the orange-brown colour remains.

The diagram opposite shows how hexane and hex-1-ene can be tested for unsaturation.

The bromine solution in the left-hand test tube stays orange-brown in colour. This means hexane must be saturated. The bromine solution in the right-hand test tube rapidly decolourises, implying that hex-1-ene is unsaturated.

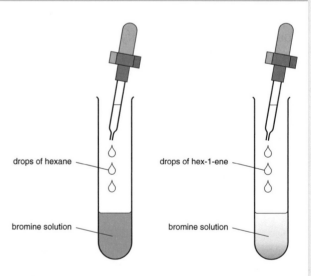

drops of hexane

drops of hex-1-ene

bromine solution

bromine solution

Explaining the test for unsaturation

When an alkene is added to bromine solution, a colour change occurs and this implies a chemical reaction has taken place. It is the presence of the C=C double bond in the alkene which allows it to react with bromine. The bromine molecule (Br-Br) adds on across the C=C double bond in the alkene. Take hex-1-ene, for example:

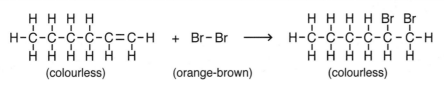

(colourless) + (orange-brown) ⟶ (colourless)

The bromine atoms become attached to those carbon atoms in the alkene that had the double bond between them. This reaction is an example of an **addition reaction**, that is one in which a small molecule adds on across the C=C double bond. In the process, the unsaturated alkene is converted into a saturated product.

More addition reactions

As well as undergoing addition reactions with bromine, alkenes can also undergo addition reactions with hydrogen and with water.

Addition of hydrogen

Consider, for example, the addition reaction between but-2-ene and hydrogen:

but-2-ene + hydrogen ⟶ butane

The hydrogen molecule adds on across the C=C double bond and butane is produced. In general, an alkene reacts with hydrogen to form the corresponding alkane. This addition reaction is sometimes referred to as **hydrogenation** because hydrogen is being added on to the alkene.

Addition of water

Let's take but-2-ene again but look at its addition reaction with water:

but-2-ene + water ⟶ butan-2-ol

A H atom from the water molecule (H–OH) adds on to one carbon atom of the double bond and the –OH left, attaches itself to the other carbon atom. Butan-2-ol is formed. In general, an alkene reacts with water to form the corresponding alkanol. This addition reaction can also be described as **hydration** since water is being added on to the alkene.

Top Tip

Alkenes undergo addition reactions with bromine, hydrogen and water to form saturated molecules that you should be able to draw and name.

Quick Test

1. What is meant by an unsaturated compound?

2. A hydrocarbon with molecular formula C_3H_6 does not rapidly decolourise bromine solution. Draw the full structural formula for this hydrocarbon and name it.

3. Name the alkene which, on reaction with bromine, would produce:

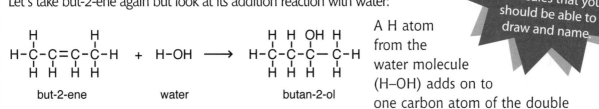

4. Name the product of the reaction between hept-3-ene and hydrogen.
5. Name the two products that form when but-1-ene reacts with water.

Answers 1. An unsaturated compound contains at least one C=C double bond. **2.** Cyclopropane
3. Pent-2-ene. **4.** Heptane. **5.** Butan-1-ol and butan-2-ol.

Cracking

Cracking in industry

Petrol is one of the more important fuels derived from crude oil and the demand for it is huge. It is made from naphtha, but the fractional distillation of crude oil does not always produce enough naphtha to satisfy demand. Fractional distillation yields more of the larger, long-chain hydrocarbons than are useful for present-day purposes. So, chemists have devised a process of converting these large, less useful hydrocarbons into smaller ones, some of which are suitable for making petrol. This process is called **cracking**.

Cracking is a chemical reaction in which alkanes are broken down to produce a mixture of smaller, more useful alkanes and alkenes:

alkanes → smaller alkanes (saturated) + smaller alkenes (unsaturated)

High temperatures are needed for cracking to take place but a **catalyst** is normally used and this allows the reaction to be carried out at a lower temperature. As a result, energy is saved and costs are reduced.

Although the original purpose of cracking was to provide more alkanes to make more petrol, it also provides alkenes which are used in many different industries, including the plastics industry.

Cracking in the laboratory

The following diagram shows how cracking can be carried out in the laboratory.

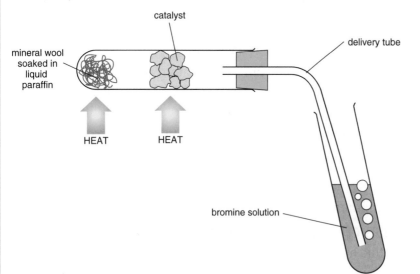

Top Tip
Remember that alkenes as well as alkanes are produced in the cracking process.

Aluminium oxide or a silicate, such as bits of broken clay pots, can be used as a catalyst. The substance being cracked is liquid paraffin, which contains large alkanes. Since the catalyst and reactant alkanes are in different states, this is an example of **heterogeneous catalysis**. The large alkanes are broken down on the surface of the catalyst and the products are passed through bromine solution. The bromine solution is rapidly decolourised and this proves that unsaturated hydrocarbons, i.e. alkenes, are among the products of cracking.

Why does cracking produce unsaturated alkenes?

When an alkane molecule is cracked, it does not react with any other substance. This means that the total numbers of carbon and hydrogen atoms in the product molecules must be equal to the numbers of carbon and hydrogen atoms in the alkane molecule that is cracked. Take hexane, for example.

$$H-\overset{\overset{H}{|}}{\underset{\underset{H}{|}}{C}}-\overset{\overset{H}{|}}{\underset{\underset{H}{|}}{C}}-\overset{\overset{H}{|}}{\underset{\underset{H}{|}}{C}}-\overset{\overset{H}{|}}{\underset{\underset{H}{|}}{C}}\overset{}{\vdots}\overset{\overset{H}{|}}{\underset{\underset{H}{|}}{C}}-\overset{\overset{H}{|}}{\underset{\underset{H}{|}}{C}}-H$$

When hexane (C_6H_{14}) breaks at the point indicated by the dotted line, butane can be formed. Since butane has the molecular formula C_4H_{10}, then the other product must have the molcular formula C_2H_4 and so must be ethene. The equation for the reaction can be written in terms of molecular or structural formulae:

$$C_6H_{14} \longrightarrow C_4H_{10} + C_2H_4$$

hexane → butane + ethene

Other bonds in the hexane molecule can break, but no matter which bond is broken one of the products must be an unsaturated alkene. There are simply not enough hydrogen atoms in an alkane molecule for it to break down into two smaller alkanes.

Cracking an alkane produces a mixture of products and the bigger the alkane, the more products will be present in this mixture.

Quick Test

1. What is meant by cracking?
2. Apart from increasing the reaction rate, why is a catalyst used in cracking?
3. Cracking can be carried out in the laboratory. Name a catalyst that can be used.
4. When hexane is cracked, ethene and another product are formed. Write an equation using molecular formulae, for this reaction and name the other product.
5. When a molecule of undecane ($C_{11}H_{24}$) was cracked, two products were obtained. If one was octane, then write the molecular formula for the other product and name it.

Answers 1. Cracking is a reaction in which alkanes are broken down into smaller, more useful, alkanes and alkenes. **2.** To allow the cracking to take place at a lower temperature and so reduce energy costs. **3.** Aluminium oxide or a silicate. **4.** $C_6H_{14} \rightarrow C_2H_4 + C_4H_{10}$: butane. **5.** C_3H_6: propene.

Ethanol

Ethanol can be made by fermenting glucose in the presence of yeast. In this **fermentation** reaction the glucose breaks down into ethanol and carbon dioxide:

$$C_6H_{12}O_6 \rightarrow 2CH_3CH_2OH + 2CO_2$$

 glucose   ethanol   carbon dioxide

An **enzyme** present in the yeast catalyses the reaction.

Fermentation can be carried out in the laboratory using the apparatus opposite:

After the glucose/yeast mixture has been left in a warm atmosphere for a few days, it is found that the limewater has turned cloudy, proving that carbon dioxide is a product of fermentation.

limewater

glucose solution plus yeast

Ethanol and alcoholic drinks

Ethanol for alcoholic drinks is made by fermenting the glucose found in any fruit or vegetable. The type of alcoholic drink produced depends on the plant source.

The fermentation process is carried out at about 27°C. When the ethanol concentration in the fermenting mixture reaches about 13% the yeast cells are killed by the excess ethanol and fermentation stops. Many alcoholic drinks, like whisky and vodka for example, have a concentration of ethanol much higher than 13%. These so-called 'spirits' are manufactured by **distilling** the fermented mixtures, and this increases the ethanol concentration. The distillation process can be demonstrated in the laboratory using the apparatus shown.

Alcoholic drink	Plant source
wine	grapes
cider	apples
whisky	barley
vodka	potatoes

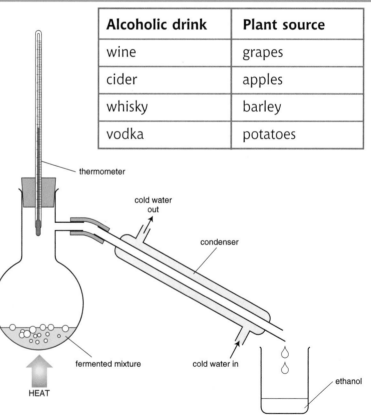

thermometer

cold water out

condenser

fermented mixture

cold water in

ethanol

HEAT

The separation of the ethanol from the water in the fermented mixture depends on the difference in the boiling points of ethanol (79°C) and water (100°C). As the mixture is heated, the ethanol boils off first since it has the lower boiling point. The ethanol vapour then passes into the **condenser** where it cools to form a liquid.

- Ethanol is used as a **solvent**.
- Ethanol is also present in **alcoholic drinks**, which can have damaging effects on the body and mind if taken in excess.
- Ethanol can also be used as a **fuel** for cars. In Brazil, for example, ethanol is made by fermenting sugar cane. It is then mixed with petrol. The big advantage ethanol has over petrol is that ethanol is a renewable energy source.

Top Tip
Remember that ethanol made by fermentation is a renewable fuel but petrol is not.

Making ethanol in industry

Ethanol made by fermentation cannot meet market demand, and so chemists have devised another method of manufacturing ethanol. This involves the reaction between ethene and water in the presence of a catalyst.

$$\underset{\text{ethene}}{\text{H}-\text{C}=\text{C}-\text{H}} \quad + \quad \underset{\text{water}}{\text{H}-\text{O}-\text{H}} \quad \longrightarrow \quad \underset{\text{ethanol}}{\text{H}-\text{C}-\text{C}-\text{H}}$$

We can recognise this as an **addition reaction** or a **hydration reaction**. Since a catalyst is used, it is more usually called **catalytic hydration**. The ethene required in this process is made by cracking the alkanes in the refinery gas fraction derived from crude oil.

Dehydrating ethanol

The catalytic hydration reaction described above is reversible, and the reverse process can be carried out in the laboratory by passing ethanol vapour over hot aluminium oxide. The ethanol is converted into ethene and water:

The oxygen and hydrogen atoms that go to form the water molecule have been shaded. Since water has been removed from the ethanol, the reaction is described as a **dehydration reaction**.

$$\text{H}-\text{C}-\text{C}-\text{H} \quad \longrightarrow \quad \text{H}-\text{C}=\text{C}-\text{H} \quad + \quad H_2O$$

Quick Test

1. Name the products of the fermentation of glucose.
2. Why is there a limit to the ethanol concentration in fermenting mixtures?
3. Give an advantage of using ethanol as a fuel when it is made from sugar cane.
4. Name the type of reaction that takes place when ethene reacts with water to form ethanol.
5. Name the type of reaction that takes place when ethanol is converted into ethene.

Answers 1. Ethanol and carbon dioxide. **2.** Because the excess ethanol kills the yeast cells. **3.** It is renewable. **4.** Catalytic hydration or hydration or addition. **5.** Dehydration.

Making and breaking esters

Making esters

Esters are made by the reaction between a **carboxylic acid** and an **alcohol**. In ester formation, the **carboxyl group** (-COOH) of the acid reacts with the **hydroxyl group** (-OH) of the alcohol. For example:

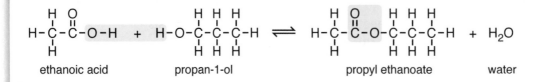

The oxygen and hydrogen atoms that go to form the water molecule have been shaded. Notice too, the shaded group of atoms in the propyl ethanoate. This is the **ester group** or **ester linkage** and identifies propyl ethanoate as an ester.

The reaction between a carboxylic acid and an alcohol is an example of a **condensation reaction**. It can also be described as **esterification** since an ester is formed.

Esters can be made in the laboratory using the apparatus drawn opposite.

paper towel soaked in cold water

hot water

alcohol and carboxylic acid plus a few drops of concentrated sulphuric acid

A few drops of concentrated sulphuric acid are added to the alcohol/carboxylic acid mixture to catalyse the reaction. The reaction rate is also speeded up by placing the test tube in hot water. The purpose of the paper towel soaked in cold water is to act as a condenser. It keeps the top of the test tube cool and so prevents the vapours of the reactants and product from escaping. After a short time, the ester is formed. It can be recognised by its characteristic 'fruity' smell.

The difference between condensation and dehydration reactions

Condensation and dehydration reactions are often confused because in both, water is a product. In a condensation reaction (see the equation above) the water molecule is removed from **two** reactant molecules whereas in a dehydration reaction (see the equation left) the water molecule is taken from just the **one** reactant molecule.

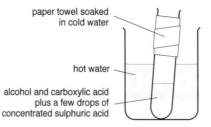

Uses of esters

Esters are used as:

- solvents
- flavourings in the food industry
- fragrances in the perfume industry.

Breaking esters

An ester can be broken down into its parent alcohol and carboxylic acid. This involves heating the ester with water in the presence of a catalyst, such as an acid or alkali. Consider the breakdown of the ester, propyl methanoate:

$$\underset{\text{propyl ethanoate}}{H-\overset{\overset{\displaystyle H}{|}}{\underset{\underset{\displaystyle H}{|}}{C}}-\overset{\overset{\displaystyle O}{\|}}{C}\vdots O-\overset{\overset{\displaystyle H}{|}}{\underset{\underset{\displaystyle H}{|}}{C}}-\overset{\overset{\displaystyle H}{|}}{\underset{\underset{\displaystyle H}{|}}{C}}-\overset{\overset{\displaystyle H}{|}}{\underset{\underset{\displaystyle H}{|}}{C}}-H} \; + \; HO-H \;\rightleftharpoons\; \underset{\text{ethanoic acid}}{H-\overset{\overset{\displaystyle H}{|}}{\underset{\underset{\displaystyle H}{|}}{C}}-\overset{\overset{\displaystyle O}{\|}}{C}-O-H} \; + \; \underset{\text{propan-1-ol}}{H-O-\overset{\overset{\displaystyle H}{|}}{\underset{\underset{\displaystyle H}{|}}{C}}-\overset{\overset{\displaystyle H}{|}}{\underset{\underset{\displaystyle H}{|}}{C}}-\overset{\overset{\displaystyle H}{|}}{\underset{\underset{\displaystyle H}{|}}{C}}-H}$$

The C-O bond in the ester link breaks and that part of the molecule on the left of the dotted line combines with the –OH group in the water molecule to form ethanoic acid. The part on the right joins with the hydrogen atom left over from the water molecule to form propan-1-ol. The breakdown of an ester by water is known as a **hydrolysis reaction**.

By looking at the equations, you will see that the hydrolysis of an ester is simply the reverse of the condensation reaction used to make the ester.

The difference between hydrolysis and hydration reactions

Both hydrolysis and hydration reactions involve water as a reactant. In hydrolysis (see the above equation), the water molecule breaks down the ester molecule into **two** product molecules but in hydration (see the equation opposite), the water molecule adds on to an alkene to give only **one** product molecule.

$$H-\overset{\overset{\displaystyle |}{C}}{\underset{\underset{\displaystyle H}{|}}{}}=\overset{\overset{\displaystyle |}{C}}{\underset{\underset{\displaystyle H}{|}}{}}-H \; + \; H_2O \;\longrightarrow\; H-\overset{\overset{\displaystyle H}{|}}{\underset{\underset{\displaystyle H}{|}}{C}}-\overset{\overset{\displaystyle O-H}{|}}{\underset{\underset{\displaystyle H}{|}}{C}}-H$$

Quick Test

1. Name the type of reaction that takes place when carboxylic acids react with alcohols to form esters.

2. Write an equation, using full structural formulae, for the reaction between methanol and propanoic acid.

3. Name the type of reaction that takes place when water reacts with an ester to form its parent alcohol and carboxylic acid.

4. Write an equation, using full structural formulae, for the hydrolysis of methyl methanoate.

Answers 1. Condensation or esterification **2.** ... **3.** Hydrolysis **4.** ...

Uses of plastics and synthetic fibres

What are plastics and synthetic fibres?

Plastics are known as **synthetic** materials because they are manufactured by the chemical industry. The vast majority are made from chemicals derived from crude oil. **Synthetic fibres** are formed when plastics are melted and squeezed through tiny holes. Examples include polyesters like Terylene and polyamides like nylon.

Properties and uses of plastics

Plastics are generally good heat and electrical insulators, and have a low density. They are resistant to most chemicals making them durable. These properties allow plastics to be used in different ways. For example:

- **PVC** is strong and rigid and used in guttering and window frames. It is an electrical insulator and used in covering electrical wires.

- **Polythene** has a low density and is strong. It is used in carrier bags and plastic bottles.

Top Tip
Make sure you are able to link the uses of plastics to their properties.

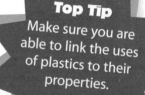

- **Polystyrene** is a heat insulator and used in hot drink cups. It is also hard and rigid and used for radio and TV casings.

- **Perspex** is transparent, hard and shatterproof. It is used in motorcycle windshields and solar panels.

- **Nylon** is strong, hard-wearing and flexible and used in tooth brush bristles and in clothes.

- **Silicones** repel water and are used as sealants in baths and showers.

Kevlar and **poly(ethenol)** are two recently developed plastics. Kevlar is very strong and used in bullet-proof vests. Poly(ethenol) is unusual in that it is soluble in water. It is used by surgeons for internal stitching.

Synthetic versus natural

For some uses, synthetic materials have advantages over natural materials and vice versa. For example, windows made from PVC rather than wood last longer and do not need to be painted. Wood, however, has the advantage that it is a renewable resource. Sweaters made from the synthetic fibre Orlon are shrink-resistant, less expensive and more hard-wearing than those made from wool. Woollen sweaters, however, are softer and warmer.

Plastics and pollution

Pollution from litter

Most plastics are **non-biodegradable**. This means that they are not broken down by bacteria, and can remain in the environment for a long time. Recently, chemists have developed some biodegradable plastics in order to reduce the litter problem. **Biopol** is one such plastic.

Pollution from burning

When plastics burn, they give off toxic gases. The gases produced depend on the elements present. All plastics contain carbon, and when they undergo incomplete combustion, **carbon monoxide** (CO) is produced. Plastics which contain hydrogen and chlorine, form **hydrogen chloride** (HCl). Plastics containing hydrogen, carbon and nitrogen, produce **hydrogen cyanide** (HCN).

Thermoplastic and thermosetting plastics

Plastics can be classified as **thermoplastic** or **thermosetting** according to how they are affected by heat. Those that are thermoplastic soften on heating and can be reshaped. Examples include polythene, polystyrene, perspex, PVC and nylon. Thermosetting plastics do not soften on heating and cannot be reshaped. Examples include bakelite and formica. They are used in electrical plugs and sockets (bakelite) and kitchen work-tops (formica).

Top Tip
Make sure you know the difference between thermoplastic and thermosetting plastics.

Quick Test

1. Most plastics are made from chemicals derived from which raw material?

2. Give one important property associated with:
   a) Kevlar   **b)** poly(ethenol)   **c)** Biopol.

3. Give one advantage that a woollen sweater has over one made from synthetic fibres.

4. Name the two toxic gases that could be formed when a plastic containing carbon, hydrogen and nitrogen is burned in a limited supply of oxygen.

5. Which type of plastic softens on heating and can be reshaped?

Answers 1. Crude oil. **2. a)** very strong **b)** soluble in water **c)** biodegradable. **3. Warmer** or softer. **4.** Carbon monoxide and hydrogen cyanide. **5.** Thermoplastic.

Addition polymerisation

Polymers

Plastics are made up of large, long-chain molecules called **polymers**. A polymer molecule is formed when hundreds of small molecules called **monomers** join together. The process of converting monomer molecules into a polymer molecule is known as **polymerisation**.

monomer molecules

polymerisation

polymer molecule

Addition polymers

The monomers used to make **addition polymers** are small unsaturated molecules derived from crude oil. Crude oil itself does not contain unsaturated molecules, but once the oil has been **fractionally distilled**, they can be produced by **cracking** some of the fractions like refinery gas and naphtha.

Let us consider **propene**, which is used to make the addition polymer called **poly(propene)**.

The propene monomer units join together by the opening of the C=C double bonds, i.e. they link up through their C=C double bonds. This is why it is important when showing the formation of an addition polymer that each monomer unit is redrawn in the shape of an '**H**' with the C=C double bond forming the crossbar of the '**H**'.

We normally show three monomer units linking together to form a section of the polymer:

To show that only a section of the polymer chain has been drawn, the end bonds must be left open. You will notice

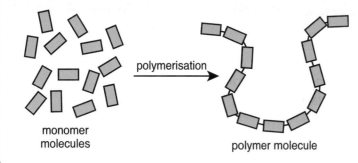

propene monomers

addition polymerisation

poly(propene)

that the backbone of poly(propene) is made up entirely of carbon atoms and this is true for all addition polymers.

Top Tip
Remember to draw the unsaturated monomer units in the shape of an 'H' to ensure that they link up through their C=C double bonds.

Naming addition polymers

The name of an addition polymer is obtained by simply writing 'poly' in front of the name of the monomer. The table shows some examples.

Monomer	Polymer
ethene	poly(ethene)
phenylethene or styrene	poly(phenylethene) or polystyrene
chloroethene or vinyl chloride	poly(chloroethene) or poly(vinyl chloride)

Identifying the monomer

Given the structure of an addition polymer, it is possible to identify its **repeating unit** and hence work out the structure of the monomer unit used to make it. Take poly(methyl methacrylate) (Perspex), for example. A section of its structure is shown below.

$$-\overset{\overset{\displaystyle H}{|}}{\underset{\underset{\displaystyle H}{|}}{C}}-\overset{\overset{\displaystyle CH_3}{|}}{\underset{\underset{\displaystyle COOCH_3}{|}}{C}}\!-\!-\!-\overset{\overset{\displaystyle H}{|}}{\underset{\underset{\displaystyle H}{|}}{C}}-\overset{\overset{\displaystyle CH_3}{|}}{\underset{\underset{\displaystyle COOCH_3}{|}}{C}}\!-\!-\!-\overset{\overset{\displaystyle H}{|}}{\underset{\underset{\displaystyle H}{|}}{C}}-\overset{\overset{\displaystyle CH_3}{|}}{\underset{\underset{\displaystyle COOCH_3}{|}}{C}}\!-\!-\!-$$

It is evident from the structure that the repeating unit is:

The end bonds in the repeating unit must be left open.

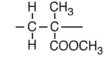

Since the monomers used to make addition polymers must be unsaturated, i.e. contain a C=C double bond, the structure of the monomer, methyl methacrylate, must be:

$$\overset{\overset{\displaystyle H}{|}}{\underset{\underset{\displaystyle H}{|}}{C}}=\overset{\overset{\displaystyle CH_3}{|}}{\underset{\underset{\displaystyle COOCH_3}{|}}{C}}$$

Quick Test

1. Name the two processes that are carried out on crude oil to obtain the monomers needed to make addition polymers.

2. Why can ethane not be used as a monomer for addition polymerisation?

3. Orlon is a synthetic fibre made from monomers with the following structure: Draw the structure of part of the Orlon polymer chain.

$$\overset{\displaystyle H}{\underset{\displaystyle H}{\diagdown}}C=C\overset{\displaystyle CN}{\underset{\displaystyle H}{\diagup}}$$

4. Name the polymer formed from but-2-ene monomer units.

5. Poly(but-1-ene) has the following structure:
   a) Draw the structure of its repeating unit.
   b) Draw the structure of its monomer unit.
   c) Name its monomer unit.

$$-\overset{\overset{\displaystyle H}{|}}{\underset{\underset{\displaystyle H}{|}}{C}}-\overset{\overset{\displaystyle CH_2CH_3}{|}}{\underset{\underset{\displaystyle H}{|}}{C}}\!-\!-\!-\overset{\overset{\displaystyle H}{|}}{\underset{\underset{\displaystyle H}{|}}{C}}-\overset{\overset{\displaystyle CH_2CH_3}{|}}{\underset{\underset{\displaystyle H}{|}}{C}}\!-\!-\!-\overset{\overset{\displaystyle H}{|}}{\underset{\underset{\displaystyle H}{|}}{C}}-\overset{\overset{\displaystyle CH_2CH_3}{|}}{\underset{\underset{\displaystyle H}{|}}{C}}\!-\!-\!-$$

Answers 1. Fractional distillation and cracking. **2.** It does not contain a C=C double bond. **3.** **4.** Poly(but-2-ene). **5. a)** **c)** but-1-ene.

Condensation polymerisation

Amines

Amines are a family of nitrogen-containing organic compounds. They can be identified by the presence of the **amine** (or **amino**) group:

$$\begin{matrix} H \\ | \\ -N-H \end{matrix} \quad \text{or} \quad -NH_2$$

Like the hydroxyl group in alcohols and the carboxyl group in carboxylic acids, the amine group is an example of a **functional group**. A functional group is the reactive part of a molecule that is involved in chemical reactions. The amine group in an amine, for example, reacts with the carboxyl group in a carboxylic acid to form an **amide link**:

$$\underset{\substack{\text{amine}\\\text{group}}}{\begin{matrix} H \\ | \\ -N-H \end{matrix}} + \underset{\substack{\text{carboxyl}\\\text{group}}}{\begin{matrix} O \\ \| \\ H-O-C- \end{matrix}} \longrightarrow \underset{\substack{\text{amide}\\\text{link}}}{\begin{matrix} H\ O \\ |\ \| \\ -N-C- \end{matrix}} + H_2O$$

Since a water molecule has been removed from two reactant molecules, the reaction to form an amide link is a **condensation reaction**.

Condensation polymers

The monomer molecules used to make condensation polymers must each contain two functional groups. With two functional groups, each monomer molecule can react with two others and so on. Only in this way can a long-chain polymer molecule be formed.

Consider nylon, a condensation polymer made from two different monomers: one a diamine and the other a dicarboxylic acid.

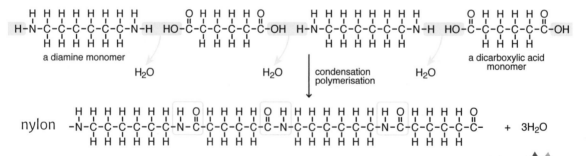

The groups circled in the nylon structure are amide links and their presence means we can describe nylon as a polyamide. You can see that the backbone of nylon contains nitrogen atoms as well as carbon atoms. In fact, all condensation polymers contain atoms of other elements as well as carbon in their backbones. This allows us to distinguish condensation polymers from addition polymers because an addition polymer has a backbone made up entirely of carbon atoms.

Top Tip
Remember that condensation polymers are made from monomers that contain two functional groups per monomer molecule.

Identifying the monomers

Given the structure of a condensation polymer it is possible to identify its repeating unit, and from this, work out the structures of the monomer units used to make it. Consider, for example, Terylene. Part of its structure is shown below.

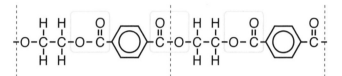

The groups circled in the structure are ester links and their presence means that Terylene can be classified as a polyester.

The repeating unit in Terylene has the following structure:

We work this out by starting at the dotted line on the left-hand side of the polymer chain. As we move along the chain we come to the second dotted line and this is the point where the structure begins to repeat itself.

But how do we use this repeating unit to work out the structures of the monomers? We know that water is a product of condensation polymerisation, and so in the reverse hydrolysis reaction, it must be a reactant. The water molecules attack the ester links and break the C-O bonds, indicated by the dotted lines. The –OH groups of the water molecules join with the carbon atoms of the C-O bonds to form carboxyl groups. The H atoms left over from the water molecules join with the oxygen atoms of the C-O bonds to form hydroxyl groups. We now have the structures of the monomer units. One is a diol and the other is a dicarboxylic acid.

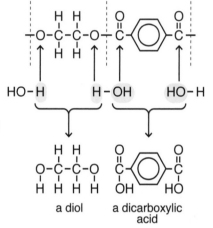

a diol   a dicarboxylic acid

Quick Test

1. Draw a structure for the functional group in amines.

2. What is formed when an amine group reacts with a carboxyl group?

3. How many functional groups must be contained in each of the monomer units used to make condensation polymers?

4. Poly(lactic acid) is a biodegradable polymer. It can be made from lactic acid, which has the following structural formula:

Draw a section of the structure of poly(lactic acid) showing three monomer units linked together.

5. Part of the structure of Kevlar is shown.

Draw structures for the two monomers used to make Kevlar.

Carbohydrates

Photosynthesis and respiration

Carbohydrates are compounds which contain the elements carbon, hydrogen and oxygen. They make up an important class of food made by plants. One of these carbohydrates called **glucose** is made in plants by a process called **photosynthesis**. The reactants in photosynthesis are carbon dioxide and water and the products are oxygen and glucose:

$$6CO_2 + 6H_2O \rightarrow C_6H_{12}O_6 + 6O_2$$
glucose

The reaction is **endothermic** and can only take place in the presence of **chlorophyll**. Its role is to absorb light energy from the **Sun**.

Carbohydrates supply the body with energy by means of a process called **respiration**. In this **exothermic** reaction, glucose reacts with oxygen to produce carbon dioxide and water:

$$C_6H_{12}O_6 + 6O_2 \rightarrow 6CO_2 + 6H_2O$$
glucose

Photosynthesis uses carbon dioxide and water and produces oxygen

O_2 CO_2 CO_2 O_2 H_2O H_2O

Top Tip
Both plants and animals respire but only plants can photosynthesise.

Classifying carbohydrates

Carbohydrates in the food we eat can be divided into two groups; **sugars** and **starch**. Sugars include **glucose**, **fructose**, **maltose** and **sucrose** (table sugar) and are made up of small molecules. Starch is a **condensation polymer**. Plants convert the glucose they make by photosynthesis, into **starch**, for storing energy.

The molecular formulae of some common carbohydrates are shown in the table opposite.

Their molecular formulae show that hydrogen and oxygen are found in carbohydrates in the same ratio as they are in water, i.e. 2:1. We can also see that glucose and fructose have the same molecular formula but since they have different structural formulae, they must be isomers. Similarly, maltose and sucrose are isomers.

Carbohydrate	Molecular formula
glucose	$C_6H_{12}O_6$
fructose	$C_6H_{12}O_6$
maltose	$C_{12}H_{22}O_{11}$
sucrose	$C_{12}H_{22}O_{11}$
starch	$(C_6H_{10}O_5)_n$

The iodine test

When brown **iodine solution** is added to starch, a blue-black colour appears.

Starch is the only compound that turns iodine solution blue-black, so this means that iodine solution can be used to test for starch.

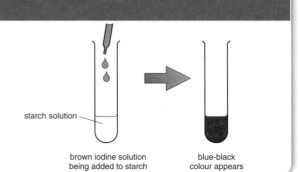

starch solution

brown iodine solution being added to starch

blue-black colour appears

Benedict's test

When some sugars, known as reducing sugars, are heated with blue **Benedict's solution**, the solution turns cloudy orange.

Of the carbohydrates mentioned in the table on the opposite page, only **glucose**, **fructose** and **maltose** turn Benedict's solution cloudy orange; sucrose and starch do not. This means that glucose, fructose and maltose are examples of reducing sugars, and can be distinguished from sucrose and starch by using Benedict's test.

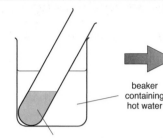
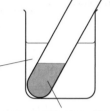

blue Benedict's solution + glucose or fructose or maltose

the reducing sugar has turned the Benedict's solution from blue to cloudy-orange

beaker containing hot water

Breaking down starch

During **digestion**, the starch present in the food we eat, reacts with water and breaks down into glucose. This process is a **hydrolysis reaction** and is catalysed by **enzymes** present in the digestive system. The glucose molecules formed are small enough to pass through the gut wall into the blood stream, which carries them to body cells where they are used in respiration and release energy.

Enzymes in the body work best at body temperature, i.e. 37°C. At lower temperatures they are less efficient and at higher temperatures they are destroyed and lose their catalytic activity, i.e. they are **denatured**.

Starch can also be hydrolysed by heating it with an **acid catalyst**.

The reaction, however, is much slower than the enzyme-catalysed hydrolysis.

Top Tip
Remember that body enzymes function best at body temperature and are destroyed at higher temperatures.

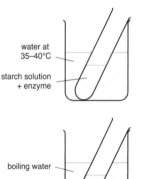

water at 35–40°C
starch solution + enzyme

boiling water
starch solution + dilute acid

Quick Test

1. Using molecular formulae, write an equation for photosynthesis.
2. Which of the following molecular formulae could represent a carbohydrate?
   a) $C_6H_{10}O_3$ **b)** $C_6H_4O_8$ **c)** $C_{18}H_{32}O_{16}$ **d)** $C_{18}H_{36}O$
3. Name the solution used to test for starch and state the colour change in this solution when it reacts with starch.
4. Which of the carbohydrates listed in the table on the opposite page will give a positive test when heated with Benedict's solution?
5. State the temperature at which body enzymes function best.

Answers 1. $6CO_2 + 6H_2O \rightarrow C_6H_{12}O_6 + 6O_2$ **2.** c) **3.** Iodine solution: brown to blue-black **4.** Glucose, fructose and maltose **5.** 37°C

Proteins

The importance of proteins

Proteins are compounds that contain the elements carbon, hydrogen, oxygen and nitrogen. They are present in all living organisms and perform a number of important functions. Proteins are the major structural materials of animal tissue, e.g. muscles, skin, hair and nails. They are also involved in the maintenance and regulation of life processes. This latter group include **enzymes** and **hormones** like insulin. **Haemoglobin** is also a member of this group of proteins.

Structure of proteins

Proteins are natural **condensation polymers** made from **amino acid monomer units**. There are about 20 different amino acids used in synthesising proteins, and the full structural formula for the simplest one is drawn below.

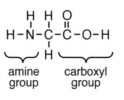

amine group   carboxyl group

Top Tip
Remember that proteins are condensation polymers made of many amino acid molecules linked together.

You will notice that it contains both an **amine (or amino) group** and a **carboxyl group** and this is true for all amino acids.

Since they each contain two **functional groups**, amino acid molecules are able to link up with each other to form a **long-chain polymer molecule**. The following equation shows three different amino acids condensing together to form part of a protein molecule.

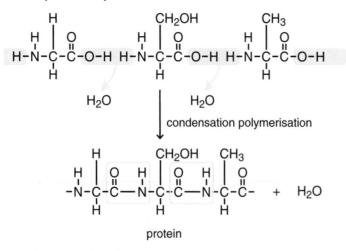

protein

You will recognise the circled groups as amide links, but when they are present in proteins, they are normally called **peptide links**. A peptide link, like an amide link, is formed by the reaction of an amine group with a carboxyl group. As mentioned above, about 20 different amino acids can be used to make proteins. So with 20 different choices available for each amino acid unit in the polymer chain, it is not surprising that there are such a vast number of different proteins.

Hydrolysis of proteins

The proteins in the food we eat are digested in the stomach and the intestines. During digestion, the proteins react with water and break down into amino acid monomer units. The process is a **hydrolysis reaction** and is catalysed by enzymes present in the digestive system. The amino acid molecules that are formed are small enough to pass through the gut wall into the bloodstream, which carries them to various sites in the body where they are reassembled into the specific proteins the body needs.

Given the structure of a section of a protein molecule, you must be able to work out the structures of the amino acid monomers that would be formed when the protein is hydrolysed. Consider for example, the following section of a protein:

During hydrolysis, the water molecules attack the peptide links and break their C-N bonds; these are indicated by the dotted lines. The –OH groups of the water molecules join with the carbon atoms of the C-N bonds to form carboxyl groups. The H atoms left over from the water molecules join with the nitrogen atoms of the C-N bonds to form amine groups. We now have the structures of the three amino acids that are formed on hydrolysing this section of the protein chain.

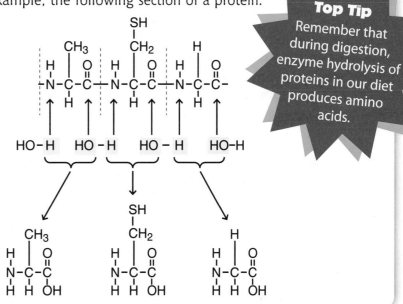

Top Tip
Remember that during digestion, enzyme hydrolysis of proteins in our diet produces amino acids.

Quick Test

1. Name the type of monomer molecules used to make proteins and name the type of polymerisation these monomers undergo.

2. Proteins contain peptide links. Name the functional groups that react to form a peptide link and name the link that is identical to the peptide link.

3. Glycine has the structural formula shown:
   Draw the full structural formula for the section of a protein made when three glycine molecules join together.

```
        H   H   O
        |   |   ||
   H  – N – C – C – O – H
            |
            H
```

Answers 1. Amino acids; condensation polymerisation. **2.** Amine group and carboxyl group; amide link. **3.**

```
         H         H         H
         |         |         |
 – C – C – N – C – C – N – C – C – N –
   ||  |       |  ||  |       |  ||  |
   O   H   H   O   H   H   O   H   H
```

Fats and oils

Fats and oils in the diet

Fats and **oils** are compounds that contain the elements, carbon, hydrogen and oxygen. Natural fats and oils can be classified as **animal**, **vegetable** or **marine** according to their origin. Some examples are given in the following table.

Animal	Vegetable	Marine
beef fat	olive oil	cod-liver oil
butter fat	linseed oil	halibut oil
mutton fat	sunflower oil	whale oil

Fats and oils are essential components of our diet. Like carbohydrates, they supply the body with energy. Carbohydrates are an immediate source of energy, and we use fats and oils as long-term energy stores. Fats and oils are a much more concentrated source of energy than carbohydrates.

Structure of fats and oils

Fats and oils are **esters** and are formed by the **condensation reaction** between the alcohol, **glycerol**, and carboxylic acids known as **fatty acids**. Glycerol has the full structural formula shown opposite.

```
    H
    |
H-C-O-H
    |
H-O-C-H
    |
H-C-O-H
    |
    H
```

Glycerol contains three hydroxyl groups (-OH) and so is classified as a **triol**.

Fatty acids are straight-chain carboxylic acids, usually with long chains of carbon atoms attached to the carboxyl group. They can be saturated or unsaturated. The structures of two typical fatty acids, stearic acid and oleic acid, are shown below.

Top Tip
You must be able to draw the structure for glycerol.

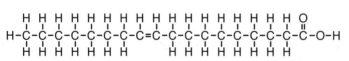

Stearic acid is saturated but oleic acid is unsaturated because it contains a C=C double bond.

Since glycerol molecules each contain three hydroxyl groups, they will react with three fatty acid molecules to form the esters present in fats and oils (R, R' and R'' represent the long chains of carbon atoms in the fatty acids).

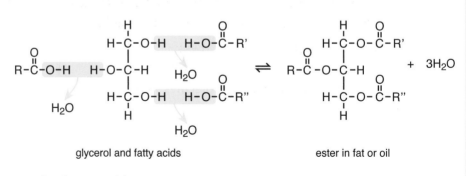

glycerol and fatty acids   ester in fat or oil

Properties of fats and oils

When samples of fats and oils are added to bromine solution, it is found that most oils rapidly decolourise it, whereas most fats have little effect. This means that the degree of unsaturation in oils is higher than it is in fats, i.e. oil molecules contain more C=C double bonds than fat molecules.

Medical evidence suggests that there is a link between the high intake of saturated fats in the diet and heart disease. This is why doctors urge us to eat less saturated fats and more unsaturated oils.

While fats are solids at room temperature, oils are liquids. This implies that the melting points of oils are lower than those of fats. The difference is related to the higher degree of unsaturation in oils.

Top Tip
Remember that the conversion of oils into hardened fats involves the partial removal of unsaturation by the addition of hydrogen.

Reactions of fats and oils

Hardening

Just as unsaturated alkenes can be converted into saturated alkanes by the addition of hydrogen, unsaturated oils can be converted into saturated fats by heating them with hydrogen in the presence of a nickel catalyst. The hydrogen molecules add on across some of the C=C double bonds in the unsaturated oil molecules, and turn them into saturated fat molecules. This process is known as hardening since it converts soft liquid oils into harder solid fats. Margarine is made in this way from corn oil and soya bean oil.

Hydrolysis

Just as a simple ester can be hydrolysed to form its parent carboxylic acid and alcohol, fats and oils react with water and break down to form fatty acids and the alcohol, glycerol.

$$
\text{fat or oil} + 3H_2O \rightleftharpoons \text{glycerol} + 3 \text{ fatty acids}
$$

Notice that 1 mole of fat or oil produces 1 mole of glycerol and 3 moles of fatty acids when hydrolysed.

Quick Test

1. Give a reason why fats and oils are an important part of a balanced diet.
2. To which family of organic compounds do fats and oils belong?
3. Name the type of acids that react with glycerol to form fats and oils.
4. Which have higher melting points, fats or oils?
5. Name the substance that reacts with oils when they are converted into hardened fats.

Answers 1. They supply the body with energy. 2. Esters. 3. Fatty acids. 4. Fats. 5. Hydrogen.

PPA 1: Testing for unsaturation

Introduction

Alkanes and cycloalkanes are saturated hydrocarbons because the carbon to carbon bonds they contain are all single bonds. Hydrocarbons, like alkenes, which contain at least one carbon to carbon double bond, are unsaturated. Orange/red bromine solution can be used to test for unsaturation. When it is added to an unsaturated hydrocarbon, it rapidly turns colourless, but when it is added to a saturated hydrocarbon, the orange/red colour remains for some time.

Aim

To test for unsaturation in four different hydrocarbons labelled $A(C_6H_{14})$, $B(C_6H_{12})$, $C(C_6H_{12})$ and $D(C_6H_{10})$, and in the light of the results suggest a possible structural formula for each one.

Procedure

Each hydrocarbon was added to a test tube to a depth of about 0.5 cm followed by about 10 drops of bromine solution. The contents were shaken by 'waggling' the test-tube.

Results

Hydrocarbon	Molecular formula	Observations on adding bromine solution	Saturated or unsaturated?
A	C_6H_{14}	remains orange/red	saturated
B	C_6H_{12}	orange/red to colourless	unsaturated
C	C_6H_{12}	remains orange/red	saturated
D	C_6H_{10}	orange/red to colourless	unsaturated

Conclusion

Hydrocarbon A is saturated and since its molecular formula is C_6H_{14} it must be an alkane. Hydrocarbon B is unsaturated and since its molecular formula is C_6H_{12} it must be an alkene. Hydrocarbon C is saturated and since its molecular formula is C_6H_{12} it must be a cycloalkane. Hydrocarbon D is unsaturated and since its molecular formula is C_6H_{10} it could be a cycloalkene or a diene. Possible structural formulae for A, B, C and D are shown below.

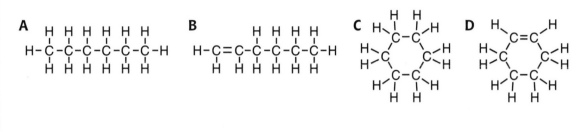

Points to note

- Bromine solution is highly **corrosive** - gloves must be worn when adding it to the hydrocarbons and any splashes on the skin should be washed off immediately with sodium thiosulphate solution.
- Since all the hydrocarbons are highly **flammable**, sources of ignition, like lighted Bunsen burners, must be absent.

Quick Test

1. How would you test a hydrocarbon for unsaturation, and what result would you get if the hydrocarbon is unsaturated?

2. Which hydrocarbon, molecular formula, C_6H_{12}, does not decolourise bromine solution?

3. Draw a possible structure for a compound with molecular formula, C_6H_{10}, which decolourises bromine solution.

4. Which two safety precautions should be taken when doing this experiment?

5. How were the contents of the test tubes shaken?

Answers 1. Use bromine solution, which would be decolourised. **2.** Cyclohexane. **3.** See structure of substance D. **4.** Wear gloves and make sure there are no naked flames. **5.** By waggling the test tube.

PPA 2: Cracking

Introduction

Cracking is an industrial process in which alkanes are split into a mixture of smaller molecules, some of which are unsaturated. Cracking is important because it converts long chain alkanes from crude oil into shorter chains for which there is greater demand. It also produces unsaturated hydrocarbons, which are important starting materials in the manufacture of plastics. If a catalyst is used in the process it can be carried out at much lower temperatures than without a catalyst, and this makes the process less expensive.

Aim

To crack liquid paraffin (a mixture of alkanes of chain length C_{20} and greater) and to find out if any of the products are unsaturated.

Procedure

The apparatus was assembled as shown in the diagram below. The catalyst was heated strongly for several seconds, and then the heat was transferred to the mineral wool to vaporise some of the liquid paraffin. The catalyst was heated again, and from time to time the heat was transferred to the mineral wool. Just before heating was stopped, the supporting clamp stand was lifted so that the delivery tube was removed from the bromine solution This was done to prevent 'suck-back'.

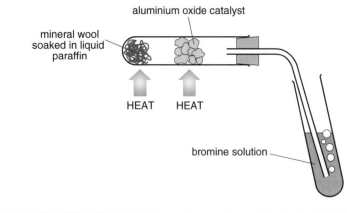

Results

The orange/red bromine solution turned colourless.

Conclusion

Since the bromine solution was decolourised, cracking must produce unsaturated hydrocarbons, i.e. hydrocarbons containing $C=C$ double bonds. For example, if $C_{22}H_{46}$ (an alkane in liquid paraffin) is cracked to produce $C_{18}H_{38}$, the molecular formula for the other product must be C_4H_8:

$$C_{22}H_{46} \rightarrow C_{18}H_{38} + C_4H_8$$

C_4H_8 is butene, which is unsaturated.

Points to note

- 'Suck-back' can occur when the pressure inside the heated test tube drops. When it occurs, the cold bromine solution is 'sucked up' the delivery tube into the hot test tube, which may break. This is why the delivery tube is lifted out of the bromine solution before heating is stopped.
- Bromine solution is highly corrosive; gloves must be worn when adding it to the hydrocarbons, and any splashes on the skin should be washed off immediately with sodium thiosulphate solution.
- Mineral wool irritates the skin and tongs must be used when handling it.

Quick Test

1. Why is cracking an important industrial process?
2. Why is a catalyst used in cracking?
3. What was the aim of this experiment?
4. Which catalyst was used in the experiment?
5. What was done to prevent suck-back?
6. What should be used to wash splashes of bromine solution from the skin?
7. What happened to the bromine solution and what can be concluded from this result?

Answers 1. Cracking converts long chain alkanes into shorter chains for which there is greater demand. It also produces unsaturated hydrocarbons used in the manufacture of plastics. **2.** So that it can be carried out at a much lower temperature. **3.** To crack liquid paraffin and find out if any products are unsaturated. **4.** Aluminium oxide. **5.** The delivery tube was removed from the bromine solution just after heating was finished. **6.** Sodium thiosulphate solution. **7.** It was decolourised so unsaturated hydrocarbons were produced.

PPA 3: Hydrolysis of starch

Introduction

Starch is a condensation polymer made from glucose monomer units. When the large starch molecules react with water they are hydrolysed (broken down) into smaller sugar molecules. We can show that the starch has been hydrolysed by heating the reaction mixture of small sugar molecules that are formed in the process with blue Benedict's solution, which will turn cloudy orange. Not all sugars give a positive Benedict's test but those formed in starch hydrolysis do.

Aim

To show that an enzyme or dilute acid catalyses the hydrolysis of starch.

Procedure (using an enzyme as catalyst)

A large beaker was half filled with water and heated to about 40°C but no higher. 3 cm^3 of starch solution was measured into each of two test tubes. 1 cm^3 of water was added to one test tube (control experiment) and 1 cm^3 of amylase solution (enzyme) was added to the other. The test tubes were placed in the beaker of warm water and left for 5 minutes. 2 cm^3 of Benedict's solution was then added to each test tube and the water heated until it boiled.

Results

Reaction mixture	Observations on heating with Benedict's solution
starch solution + water (control)	no change in colour
starch solution + amylase	blue Benedict's changed to cloudy orange

Conclusion

The positive Benedict's test shows that the starch has been hydrolysed using amylase. The negative Benedict's test in the control proves that amylase catalyses the hydrolysis.

Procedure (using an acid as catalyst)

10 cm³ of starch solution was added to each of two small beakers. 1 cm³ of water was added to one beaker (control experiment) and 1 cm³ of dilute hydrochloric acid was added to the other. The mixtures were boiled gently for 5 minutes. Sodium hydrogencarbonate was added to the beaker containing the acid/starch mixture until no more bubbles of gas were produced. 5 cm³ of Benedict's solution was added to each beaker and the reaction mixtures were heated.

Results

Reaction mixture	Observations on heating with Benedict's solution
starch solution + water (control)	no change in colour
starch solution + acid	blue Benedict's changed to cloudy orange

Conclusion

The positive Benedict's test shows that the starch has been hydrolysed using acid. The negative Benedict's test in the control proves that acid catalyses the hydrolysis.

Points to note

- The control was necessary to show the catalytic effect of the enzyme or acid on the hydrolysis of starch.
- After heating the starch solution/acid mixture, sodium hydrogencarbonate had to be added to neutralise the acid since Benedict's test does not work in acidic conditions. The sodium hydrogencarbonate has to be added in tiny amounts, since its reaction with acid can be quite violent.

Quick Test

1. Which two substances can be used to hydrolyse starch?
2. At the end of the experiment what colour change is undergone by the Benedict's solution?
3. What is added to the starch solution in the control experiment and why is the control experiment necessary?
4. Why is sodium hydrogencarbonate added to the starch solution/acid mixture?
5. Why should the sodium hydrogencarbonate solution be added in tiny amounts?

Answers 1. Amylase (an enzyme) and an acid. **2.** From blue to orange. **3.** Water. The control was necessary to show that the hydrolysis needed the enzyme or the acid. **4.** To neutralise the acid, since Benedict's solution does not work in acid. **5.** Because its reaction with acid can be quite violent.

The pH scale

pH scale and indicators

The pH scale is a continuous scale which runs from below 0 to above 14, but the pH values of most solutions you work with lie between 1 and 14.

- Acids have pH values below 7.
- Alkalis have pH values above 7
- Neutral solutions such as pure water have pH = 7.

Indicators are used to show whether a solution is acidic, alkaline or neutral. An indicator will be one colour in an acid and a different colour in an alkali. There are many different indicators but universal indicator is the one you will use most. Universal indicator is red in acid, green when it is neutral and blue/purple in an alkali. The pH of a substance can also be tested using moist pH paper. pH paper turns the same colours as universal indicator.

Top Tip
Remember that the lower the pH value the more acidic is the solution. Alkalis have higher pH values.

Non-metal oxides and metal oxides

The Periodic Table is separated into metals and non-metals by a zigzag line.

Non-metal oxides

Non-metal oxides that are soluble in water dissolve to produce **acid solutions**. Examples include:

- CO_2 dissolves to give carbonic acid, H_2CO_3.
- SO_2 dissolves to give sulphurous acid, H_2SO_3, present in acid rain.

One non-metal oxide that is neutral is H_2O.

Metal oxides

Metal oxides that are soluble in water dissolve to form **alkaline solutions**. Examples include:

- Na_2O dissolves to give sodium hydroxide, NaOH.
- CaO dissolves to give calcium hydroxide, $Ca(OH)_2$, also known as limewater.

Common acids and alkalis

Alkalis have a pH greater than 7. Common laboratory alkalis include:
- sodium hydroxide, NaOH   • ammonia, NH_3 • limewater, $Ca(OH)_2$

Common household alkalis include:
- baking soda   • oven cleaner   • dishwashing powder   • bleach

Acids have a pH less than 7. Common laboratory acids are:
- hydrochloric acid, HCl   • sulphuric acid, H_2SO_4 • nitric acid, HNO_3

Common household acids include:
- vinegar   • fizzy drinks   • fruit juices.

Both acids and alkalis are corrosive.

CORROSIVE

Equilibrium in aqueous solutions

A small proportion of water molecules break up into hydrogen ions, H^+, and hydroxide ions, OH^-, ions. Some of these hydrogen and hydroxide ions then react together again to form water molecules.

$H_2O(l) \rightarrow H^+(aq) + OH^-(aq)$ and $H^+(aq) + OH^-(aq) \rightarrow H_2O(l)$

Since it is the same reaction going in opposite directions it is known as a reversible reaction. When the **rate of the reverse reaction is the same as the rate of the forward reaction** the reaction is at **equilibrium**.

The equation is $H_2O(l) \rightleftharpoons H^+(aq) + OH^-(aq)$ where $\rightleftharpoons$ indicates that the reaction is at equilibrium.

This equilibrium is present in water and all aqueous solutions. At **equilibrium**, the **concentrations of the reactants and products remain constant** but not necessarily equal.

In water and neutral solutions, the concentration of hydrogen ions is equal to the concentration of hydroxide ions.

We can write this as $[H^+] = [OH^-]$ where [ ] means concentration of.

All acidic solutions contain more hydrogen ions than hydroxide ions.

All alkaline solutions contain more hydroxide ions than hydrogen ions.

The pH scale is a measure of the **concentration of hydrogen ions**, $[H^+]$ in a solution. The **more hydrogen ions** that are present, the **more acidic** the solution is and the **lower the pH** value.

Top Tip
Try to remember
Neutral, $[H^+] = [OH^-]$
Acid, $[H^+] > [OH^-]$
Alkaline, $[OH^-] > [H^+]$

Adding water decreases the concentration of acids and alkalis.

When acids are diluted the pH increases towards 7 as the solution becomes neutral. The concentration of the hydrogen ions decreases until $[H^+] = [OH^-]$ at pH = 7. The pH will remain at 7 even if more water is added.

When alkalis are diluted the pH decreases towards 7 as the solution becomes neutral. The concentration of the hydroxide ions decreases until $[OH^-] = [H^+]$ at pH = 7. The pH will remain at 7 even if more water is added.

Quick Test

1. An oxide dissolves in water to give a solution of pH 3. Is this likely to be a metal oxide or non-metal oxide?

2. In a solution of pH 3, what will be the relative concentrations of hydrogen and hydroxide ions?

3. Write a balanced equation for sodium oxide dissolving in water to form sodium hydroxide.

4. Write the equation that represents the equilibrium present in water.

5. What is meant when we say a reaction is at equilibrium?

Answers 1. Non-metal oxide. **2.** $[H^+] > [OH^-]$. **3.** $Na_2O + H_2O \rightarrow 2NaOH$. **4.** $H_2O \rightleftharpoons H^+ + OH^-$. **5.** The rates of the forward and reverse reactions are equal.

Concentration

Concentration of a solution

The concentration of a solution is the **number of moles of solute dissolved in a certain volume of solution**. In chemistry, the concentration is **measured in units of moles per litre (mol l^{-1})**. For example:

1 mole of sodium hydroxide, NaOH, has a mass of 40.0 g.

So 1 litre of 1 mol l^{-1} NaOH(aq) will contain 40.0 g of sodium hydroxide in 1 litre of solution.

Another triangle, which is useful in calculations involving volume and concentration of solutions, such as dilute acids and alkalis, is shown below.

Top Tip
Make sure you know and can use this formula triangle.

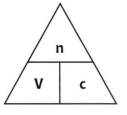

n = number of moles	Using this triangle,
V = volume of solution (**in litres**)	n = V × c
c = concentration of solution (**in mol l^{-1}**)	c = n/V

Example

Question 1: Calculate the number of moles in 500 cm^3 of a 0.2 mol l^{-1} solution.

Worked answer: The volume, V, is 500 cm^3 = **0.5 litres**
The concentration, c = **0.2 mol l^{-1}**
n = V × c = **0.5 × 0.2 = 0.1 moles**

Concentration calculations using both formula triangles

On page 34 and 35 of this book you did some calculations on 'moles to mass and mass to moles' using another formula triangle. Sometimes, in more difficult calculations, it is necessary to use both triangles.

Examples

When preparing solutions of a certain concentration, it is necessary to work out how much of the solute is needed.

Question 1: Calculate the mass of sodium hydroxide required to prepare 250 cm^3 of a solution of concentration 0.1 mol l^{-1}.

Worked answer: In this calculation we are told:
- the volume, V = 250 cm^3 which is 0.25 litres and
- the concentration, c = 0.1 mol^{-1}

We can **calculate the number of moles** using the formula

n = V × c = 0.25 × 0.1 = **0.025 moles**

The formula of sodium hydroxide is NaOH, so the formula mass can be calculated as: FM = 23 + 16 + 1 = 40.

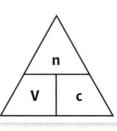

To calculate the **mass** we have to use the other formula triangle

Mass = n × FM = 0.025 × 40 = **1.0 g**

So the mass of sodium hydroxide required is **1.0 g**

Question 2: What will be the concentration of 500 cm³ of solution containing 50.0 g of sodium hydroxide?

Worked answer: In this calculation we are told:
- the volume, V = 500 cm³ which is 0.50 litres and
- the mass which is 50.0 g of NaOH.

We can work out the formula mass of NaOH to be 40.0, so we can work out the number of moles of NaOH using this triangle.

The number of moles, n = mass/FM = 50.0/40.0 = **1.25 moles**

The concentration can now be calculated using the other triangle.

c = n / V = 1.25 / 0.5 = **2.5 mol l⁻¹**

So the concentration of the sodium hydroxide solution will be **2.5 mol l⁻¹**.

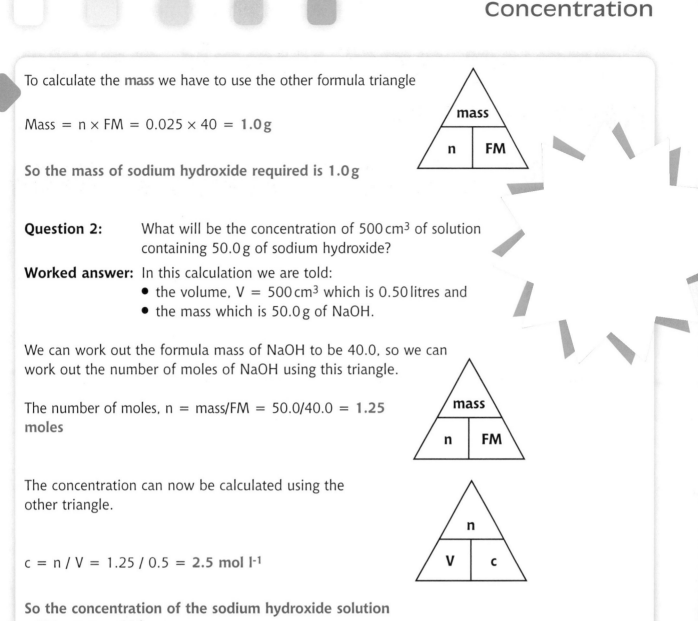

Quick Test

1. Calculate the number of moles in:
a) 100 cm³ of 0.2 mol l⁻¹ solution
b) 50 cm³ of 0.25 mol l⁻¹ solution
c) 250 cm³ of 1.0 mol l⁻¹ solution
d) 100 cm³ of 0.04 mol l⁻¹ solution.

2. Calculate the mass of sodium hydroxide required to prepare:
a) 100 cm³ of 0.2 mol l⁻¹ solution
b) 2 litres of 0.015 mol l⁻¹ solution
c) 50 cm³ of 0.02 mol l⁻¹ solution
d) 250 cm³ of 2.5 mol l⁻¹ solution.

3. Calculate the mass of sodium nitrate required to prepare:
a) 100 cm³ of 0.2 mol l⁻¹ solution
b) 2 litres of 0.01 mol l⁻¹ solution
c) 50 cm³ of 0.2 mol l⁻¹ solution
d) 250 cm³ of 0.1 mol l⁻¹ solution

4. What will be the concentration of 200 cm³ of a solution containing:
a) 4.8 g of LiOH?
b) 2.0 g of NaOH?
c) 2.0 g of calcium hydroxide?
d) 0.85 g of silver(I) nitrate?

Answers 1. a) 0.02 moles **b)** 0.0125 moles **c)** 0.25 moles **d)** 0.004 moles. **2. a)** 0.8 g **b)** 1.2 g **c)** 0.040 g **d)** 25.0 g. **3. a)** 1.7 g **b)** 1.7 g **c)** 0.85 g **d)** 2.125 g.
4. a) 1.0 mol l⁻¹ **b)** 0.25 mol l⁻¹ **c)** 0.135 mol l⁻¹ **d)** 0.025 mol l⁻¹.

Strong and weak acids and bases

Strong and weak acids

In chemistry, the words strong and weak do not mean the same as concentrated and dilute. A concentrated solution contains much more solute dissolved in a certain volume of solution compared to a dilute solution. A strong acid can be either concentrated or dilute, depending on the number of moles of acid per litre of solution. The same applies to a weak acid.

In aqueous solution, strong acids are completely dissociated into ions but weak acids are only partially dissociated. This means that when a strong acid is made into an aqueous solution all the molecules break up into ions.

- Hydrochloric acid is a strong acid and it dissociates completely in water:
  $HCl(aq) \rightarrow H^+(aq) + Cl^-(aq)$
- Nitric acid, HNO_3, and sulphuric acid, H_2SO_4, are also strong acids.
- Ethanoic acid, CH_3COOH, is a weak acid and it is only partially dissociated in water. This means that only a small number of ethanoic acid molecules break up into ions when in aqueous solution. This can be shown in the equation:
  $CH_3COOH(aq) \rightleftharpoons CH_3COO^-(aq) + H^+(aq)$

  | ethanoic acid | ethanoate | hydrogen |
  | molecules | ions | ions |

The double arrow shows that not all the ethanoic acid molecules turn into ions.

Top Tip
The formula of the ethanoate ion is given on page 4 of your Data Booklet.

The table below shows the results of some experiments using 0.1 mol l⁻¹ hydrochloric acid (a strong acid) and 0.1 mol l⁻¹ ethanoic acid (a weak acid).

Acid	pH	Conductivity/ mA	Rate of reaction with Mg
Hydrochloric acid, HCl	1.0	94	Hydrogen gas produced very quickly
Ethanoic acid, CH_3COOH	2.9	6	Hydrogen gas produced very slowly

These results tell us that there are more ions in hydrochloric acid compared to ethanoic acid, despite the fact that they are the same concentration (equimolar).

Strong and weak bases

In aqueous solution, strong bases are completely ionised but weak bases are only partially ionised.

Strong bases

Solutions of metal hydroxides are strong bases. Not all metal hydroxides are soluble in water. The hydroxides of the Group 1 metals are very soluble (see page 5 in the Data Booklet). So lithium hydroxide, sodium hydroxide and potassium hydroxide are strong bases which dissolve to form alkaline solutions.

$$Na^+OH^-(s) + H_2O(l) \rightarrow Na^+(aq) + OH^-(aq)$$

Hydroxides of some of the Group 2 elements are soluble in water but the hydroxides of most metals are insoluble in water. Metal oxides and metal hydroxides have ionic bonding but are only alkaline if they are soluble in water:

Top Tip
Don't get mixed up between strong and concentrated or weak and dilute.

Weak bases

Ammonia, NH_3, has polar covalent bonding and is very soluble in water. Some of the ammonia molecules react with the water molecules to form ammonium ions and hydroxide ions. Most of the ammonia molecules are not ionised and so ammonia is a weak base:

$$NH_3(g) + H_2O(l) \rightleftharpoons NH_4^+(aq) + OH^-(aq)$$

Just as with strong and weak acids, strong and weak bases differ in pH and conductivity as can be seen in the table below.

Base	Concentration	pH	Conductivity/mA
Sodium hydroxide, NaOH	0.1 mol l⁻¹	13.0	66
Ammonia, NH_3	0.1 mol l⁻¹	11.1	4

These results show that the 0.1 mol l⁻¹ sodium hydroxide solution contains many more ions than the 0.1 mol l⁻¹ solution and that sodium hydroxide is a strong base and ammonia is a weak base.

Quick Test

1. What is meant by a strong acid?
2. What is meant by a weak base?
3. Name and give the formulae of three strong acids.
4. Name and give the formula of a weak acid.
5. Name and give the formulae of three strong bases.

6. Write an equation for ammonia dissolving in water.
7. Explain why 1 mol l⁻¹ hydrochloric acid solution has a lower pH than 1 mol l⁻¹ ethanoic acid solution.

Reactions of acids

Neutralisation

Neutralisation is the reaction of **acids** and **bases**. Metal oxides, metal hydroxides and metal carbonates are examples of bases. Soluble bases dissolve in water to form alkalis. In a neutralisation reaction the pH of an acid rises towards 7 and the pH of an alkali falls down towards 7. During a neutralisation reaction, water and a **salt** are always formed.

Everyday examples of neutralisation include:

- putting vinegar onto the alkaline sting from a wasp
- putting baking soda solution onto the acid sting from a bee
- using lime (CaO) to reduce acidity in soil and lochs
- taking antacid tablets to treat acid indigestion.

Acids and alkalis

Alkalis neutralise acids forming a salt and water.

The formula equation for sodium hydroxide neutralising nitric acid to form the salt, sodium nitrate and water is
$$HNO_3(aq) + NaOH(aq) \rightarrow NaNO_3(aq) + H_2O(l)$$

In the reaction of any acid with any alkali, the H^+ ions from the acid react with the OH^- ions from the alkali to form water:
$$H^+(aq) + OH^-(aq) \rightarrow H_2O(l)$$

Top Tip
- Nitric acid forms nitrate salts.
- Hydrochloric acid forms chloride salts.
- Sulphuric acid forms sulphate salts.

Acids and metal oxides

Metal oxides neutralise acids forming a salt and water.

The formula equation for copper(II) oxide reacting with sulphuric acid to form copper(II) sulphate and water is
$$H_2SO_4(aq) + CuO(s) \rightarrow CuSO_4(aq) + H_2O(l)$$

In the reaction of any acid and any metal oxide, the H^+ ions from the acid react with the oxide ions to form water.
$$2H^+(aq) + O^{2-}(s) \rightarrow H_2O(l)$$

Top Tip
When an acid reacts it is the H^+ ions from the acid that change. The negative ion in an acid usually stays unchanged.

Acids and metal carbonates

Metal carbonates neutralise acids forming a salt, water and carbon dioxide.

In the reaction the H^+ ions from the acid are reacting with the CO_3^{2-} ions from the metal carbonate to form water and carbon dioxide:
$$2H^+ + CO_3^{2-} \rightarrow H_2O + CO_2$$

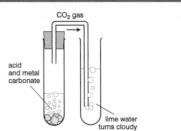

CO₂ gas

acid and metal carbonate

lime water turns cloudy

Acids and metals

Acids react with some metals to form a salt and hydrogen gas.

The hydrogen produced burns with a squeaky pop. This is the test for hydrogen gas.

The word equation for magnesium reacting with sulphuric acid is:
sulphuric acid + magnesium → magnesium sulphate + hydrogen

The formula equation is:
$H_2SO_4(aq) + Mg(s) \rightarrow MgSO_4(aq) + H_2(g)$

The hydrogen ions from the acid change into molecules of hydrogen gas.

Top Tip
You should know the formulae of the different acids, the products of the neutralisation reactions, and be able to write formulae equations for the reactions.

'squeaky' pop

H_2 gas

burning splint

Mg and H_2SO_4

Acid rain

When fossil fuels are burned **sulphur dioxide** may be produced from sulphur impurities. The spark that ignites the petrol/air mixture in a car engine has enough energy to change nitrogen and oxygen in the air into **nitrogen dioxide**. Both nitrogen dioxide and sulphur dioxide dissolve in water in the atmosphere to produce **acid rain**.

Acid rain damages:

- buildings made from carbonate rocks by eroding them
- structures made from iron and steel by making them corrode/rust faster
- plant life by making the soil too acidic for plants to grow
- animal life, particularly life in ponds, lochs and lakes by making the water too acidic.

Quick Test

1. Write a balanced chemical equation for the reaction between hydrochloric acid and potassium hydroxide.

2. Which ions in an acid always react when the acid is being neutralised?

3. What are the products when zinc oxide reacts with sulphuric acid?

4. Write the balanced chemical equation for calcium carbonate reacting with nitric acid.

5. Write the chemical formula for two gases that contribute to acid rain.

Answers 1. $HCl + KOH \rightarrow KCl + H_2O$ **2.** Hydrogen ions, H^+ **3.** Zinc sulphate and water. **4.** $CaCO_3 + 2HNO_3 \rightarrow Ca(NO_3)_2 + 2H_2O + CO_2$ **5.** NO_2 and SO_2

Volumetric titrations

What is a titration?

A **titration** experiment involves using a **burette**, which is used to dispense accurately measured volumes of a liquid, usually the acid solution. The acid from the burette flows into a **conical flask**, which would normally contain a known volume of the alkali plus a few drops of **indicator**. The indicator changes colour when the **neutralisation reaction** is just complete. The volume of acid used can then be read off the scale on the side of the burette. Titrations are usually repeated until two **concordant results** (within 0.1 cm³ of each other) are obtained. This makes the results reliable.

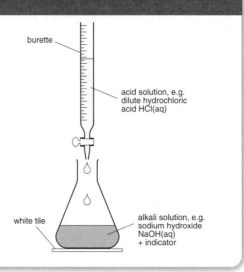

burette

acid solution, e.g. dilute hydrochloric acid HCl(aq)

white tile

alkali solution, e.g. sodium hydroxide NaOH(aq) + indicator

Calculation of concentration from results of titrations

Titrations are usually carried out to determine the **concentration** of the acid or the concentration of the alkali.

If the volumes of both the acid and alkali are known and the concentration of one of them also known, then the concentration of the other can be calculated.

To do this we can use the formula:

$P_{acid} \times V_{acid} \times c_{acid} = P_{alkali} \times V_{alkali} \times c_{alkali}$

Where: P = power V = volume c = concentration

- **The power of an acid is the number of H⁺ ions in the formula**
  (e.g. hydrochloric acid HCl has a power of 1; sulphuric acid H_2SO_4 has a power of 2).

- **The power of an alkali is the number of OH⁻ ions in the formula**
  (e.g. sodium hydroxide NaOH has a power of 1; calcium hydroxide $Ca(OH)_2$ has a power of 2).

Example

Question: In a titration experiment, 16.2 cm³ of 0.1 mol l⁻¹ sulphuric acid was needed to neutralise 25.0 cm³ of sodium hydroxide solution. Calculate the concentration of the sodium hydroxide solution.

Worked Answer: $P_{acid} \times V_{acid} \times c_{acid} = P_{alkali} \times V_{alkali} \times c_{alkali}$

$$2 \times 16.2 \times 0.1 = 1 \times 25.0 \times c_{alkali}$$

$$\text{So } c_{alkali} = \frac{2 \times 16.2 \times 0.1}{1 \times 25.0}$$

$$= 0.13 \text{ mol l}^{-1}$$

So the concentration of the sodium hydroxide was **0.13 mol l⁻¹**

Calculating volume of acid or alkali required for neutralisation

If you know the concentration of both the acid and the alkali and the volume of one of them, you should also be able to calculate the other volume required for complete neutralisation to take place. The same formula is used as before, but this time the unknown is either the volume of the acid or the volume of the alkali.

Example

Question: Calculate the volume of 0.10 mol l^{-1} hydrochloric acid required to exactly neutralise 20.0 cm^3 of 0.04 mol l^{-1} calcium hydroxide solution.

Worked answer: HCl has power = 1, so P_{acid} = 1
The acid has concentration 0.10 mol l^{-1}, so c_{acid} = 0.10
The unknown is V_{acid}
Ca(OH)$_2$ has power = 2, so P_{alkali} = 2
The volume of the alkali is 20.0 cm^3, so V_{alkali} = 20.0
The calcium hydroxide solution has concentration 0.04 mol l^{-1}, so c_{alkali} = 0.04

$$P_{acid} \times V_{acid} \times c_{acid} = P_{alkali} \times V_{alkali} \times c_{alkali}$$
$$1 \times V_{acid} \times 0.10 = 2 \times 20.0 \times 0.04$$
$$\text{So } V_{acid} = \frac{2 \times 20.0 \times 0.04}{1 \times 0.10}$$
$$= 16.0 \text{ cm}^3$$

So the volume of hydrochloric acid required is 16.0 cm^3.

Top Tip
Remember that the same formula is used to calculate the volume required as to calculate the concentration – it is just that the unknown is different.

Quick Test

1. 20.0 cm^3 of 0.1 mol l^{-1} sodium hydroxide solution was exactly neutralised by 10.0 cm^3 of hydrochloric acid. Calculate the concentration of the hydrochloric acid.

2. 20.0 cm^3 of 0.1 mol l^{-1} sodium hydroxide solution was exactly neutralised by 20.0 cm^3 of sulphuric acid. Calculate the concentration of the sulphuric acid.

3. 25.0 cm^3 of 1.0 mol l^{-1} sodium hydroxide solution was exactly neutralised by 18.4 cm^3 of nitric acid. Calculate the concentration of the nitric acid.

4. 20.0 cm^3 of 0.02 mol l^{-1} sulphuric acid was exactly neutralised by 15.8 cm^3 of potassium hydroxide solution. Calculate the concentration of the potassium hydroxide solution.

5. What volume of 0.12 mol l^{-1} hydrochloric acid will neutralise:
   a) 25.0 cm^3 of 0.05 mol l^{-1} calcium hydroxide solution?
   b) 20.0 cm^3 of 0.20 mol l^{-1} sodium hydroxide solution?

Answers 1. 0.2 mol l^{-1} **2.** 0.05 mol l^{-1} **3.** 1.4 mol l^{-1} **4.** 0.051 mol l^{-1} **5. a)** 20.83 cm^3 **b)** 33.33 cm^3

Naming salts

Names of salts produced

A **salt** is a compound in which the hydrogen ions of an acid have been replaced by metal ions (or ammonium ions). When an acid reacts with a base (such as an alkali, metal oxide or metal carbonate) or with a metal, a salt is produced. The name of the salt produced depends on the acid used and the base or metal it reacts with.

Neutralising **hydrochloric** acid produces **chloride** salts:
sodium hydroxide + hydro**chloric** acid → **sodium chloride** + water

Neutralising **nitric** acid produces **nitrate** salts:
calcium oxide + **nitric** acid → **calcium nitrate** + water

Neutralising **sulphuric** acid produces **sulphate** salts:
copper(II) carbonate + **sulphuric** acid → **copper(II) sulphate** + water + carbon dioxide

Nitrogen salts

Both animals and plants require compounds of nitrogen for healthy growth. Animals get nitrogen compounds from their foods but plants usually take in nitrogen compounds through their roots. These nitrogen compounds must be soluble in water for this to happen. All ammonium salts and nitrate salts are soluble in water. Some nitrogen salts, including ammonium nitrate, ammonium sulphate and potassium nitrate are made by neutralisation reactions for use as fertilisers (as they provide compounds of nitrogen and are soluble).

Top Tip
The formula for the ammonium ion is given on page 4 of your Data Booklet.

Preparation of soluble salts

Soluble salts are prepared from the appropriate acid.

Acid and alkali

The acid and alkali are mixed together until the solution is neutral. This can be checked with pH paper. The water in the salt solution formed is then evaporated off leaving the solid salt behind. This is a good method of preparing sodium and potassium salts.

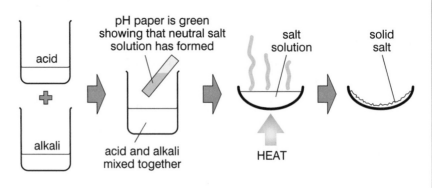

acid
+
alkali

acid and alkali mixed together

pH paper is green showing that neutral salt solution has formed

salt solution

HEAT

solid salt

Other methods involve reacting the appropriate acid with a solid such as a metal, metal oxide or metal carbonate. The general method is to add excess of the solid to make sure that all the acid has been used up. The solid left over is then filtered off and the salt solution passes through the filter paper. The solid salt is then isolated by evaporating off the water.

Acid and metal

Metals can be reacted with acids to form a salt and hydrogen:

metal + acid → salt + hydrogen

zinc + hydrochloric acid → zinc chloride + hydrogen

$Zn(s) + 2HCl(aq) \rightarrow ZnCl_2(aq) + H_2(g)$

magnesium + sulphuric acid → magnesium sulphate + hydrogen

$Mg(s) + H_2SO_4(aq) \rightarrow MgSO_4(aq) + H_2(g)$

In the preparation of soluble salts it is often easier to use an insoluble metal carbonate or metal oxide as the base.

Acid and metal oxide

Metal oxides are bases; they can be reacted with acids to make salts and water:

metal oxide + acid → salt + water

copper oxide + hydrochloric acid → copper chloride + water

$CuO(s) + 2HCl(aq) \rightarrow CuCl_2(aq) + H_2O(l)$

zinc oxide + sulphuric acid → zinc sulphate + water

$ZnO(s) + H_2SO_4(aq) \rightarrow ZnSO_4(aq) + H_2O(l)$

Acid and metal carbonate

Metal carbonates are bases; they can be reacted with acids to make salts.

Copper chloride can be made using hydrochloric acid and copper carbonate:

copper carbonate + hydrochloric acid → copper chloride + water + carbon dioxide

$CuCO_3(s) + 2HCl(aq) \rightarrow CuCl_2(aq) + H_2O(l) + CO_2(g)$

- Copper carbonate is added to the acid until it stops fizzing.
- The unreacted copper carbonate is then removed by **filtering**.
- The solution is poured into an evaporating dish.
- It is heated until the first crystals appear.
- The solution is then left for a few days for the copper chloride to **crystallise**.

Top Tip

When preparing a soluble salt by reacting a solid metal, metal oxide or metal carbonate with an acid, the solid must be added in excess to make sure that all the acid is used up. The excess solid is then filtered and the salt solution is heated to evaporate off the water, leaving behind the solid salt.

Quick Test

1. Which acid and alkali would be used to prepare sodium nitrate?

2. Which acid and metal would be used to prepare zinc sulphate?

3. Which acid and metal carbonate would be used to prepare copper(II) sulphate?

4. Which acid and metal oxide would be used to prepare copper(II) nitrate?

5. Write balanced chemical equations for the above four reactions.

Answers 1. Sodium hydroxide and nitric acid. **2.** Zinc and sulphuric acid. **3.** Copper(II) carbonate and sulphuric acid. **4.** Copper(II) oxide and nitric acid. **5.** $HNO_3 + NaOH \rightarrow NaNO_3 + H_2O$; $H_2SO_4 + Zn \rightarrow ZnSO_4 + H_2$; $H_2SO_4 + CuCO_3 \rightarrow CuSO_4 + H_2O + CO_2$; $2HNO_3 + CuO \rightarrow Cu(NO_3)_2 + H_2O$

Precipitation and formulae of ions

What is precipitation?

When two solutions react to form an **insoluble** product, the insoluble product is the **precipitate** and it is known as a **precipitation** reaction.

Preparing insoluble salts

When asked to prepare a salt, it is necessary to find out first if the salt is **soluble** or **insoluble**. This can be done by looking at page 5 of the Data Booklet.

If the salt to be prepared is **insoluble**, then it must be prepared by **precipitation**. This is done by preparing one solution containing the positive ions in the salt and another solution containing the negative ions in the salt. The two solutions are then mixed together to form a precipitate of the salt which is then filtered off.

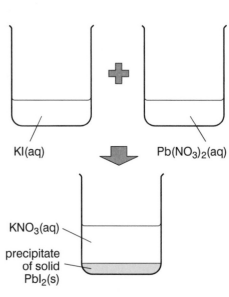

For example, insoluble lead iodide is made by preparing a solution containing lead ions, such as **lead** nitrate, and another solution containing iodide ions, such as potassium **iodide**. When these two solutions are mixed together a precipitate of **lead iodide** immediately forms:

$Pb(NO_3)_2(aq) + 2KI(aq) \rightarrow PbI_2(s) + 2KNO_3(aq)$

Another insoluble salt is silver chloride. It has formula AgCl and contains positive silver ions and negative chloride ions. To prepare silver chloride by precipitation, a solution containing silver ions must be mixed with a solution containing chloride ions.

Looking at page 5 in the Data Booklet, we can see that suitable compounds would be silver nitrate and sodium chloride since they are both soluble.

Solutions of silver nitrate and sodium chloride would be prepared and when mixed together the white precipitate of silver chloride would form immediately.

Formulae of ions

Metallic elements form **positive** ions. **Non-metal** elements usually form **negative** ions. The size of the charge depends on the **valency** of the element. For example:

- Sodium is a metal in Group 1 and so has valency=1. Sodium ions will have a charge 1+ and the formula of the sodium ion is Na^+

- Magnesium is a metal and has valency 2 so magnesium ions have charge 2+ and the formula of the magnesium ion is Mg^{2+}.

- Aluminium is a metal and has valency 3 so aluminium ions have charge 3+ and the formula of the aluminium ion is Al^{3+}.

- Chlorine is a non-metal in Group 7 and has valency 1 so chloride ions have a charge of 1- and the formula of the chloride ion is Cl^-.

- Sulphur is a non-metal in Group 6 and has valency = 2 so sulphide ions have a charge of 2- and the formula of the sulphide ion is S^{2-}

- Copper(II) ions have formula Cu^{2+} and copper(I) ions have formula Cu^+.

- Iron(III) ions have formula Fe^{3+}.

- The formulae of ions containing more than one atom such as sulphate, nitrate etc are given on page 4 of the Data Booklet.

Quick Test

1. What is meant by a precipitate?

2. What solution should be mixed with sodium iodide to form a precipitate of silver iodide?

3. What would be the precipitate formed when a solution of lead nitrate is mixed with a solution of potassium carbonate?

4. Write ionic formulae for the:
   a) the aluminium ion b) the oxide ion c) the calcium ion d) lithium ion
   e) fluoride ion f) nickel(II) ion g) titanium(III) ion h) sulphate ion i) hydroxide ion
   j) carbonate ion k) nitrate ion l) ethanoate ion m) phosphate ion

Answers 1. A solid that is formed when two solutions are mixed together. **2.** Silver nitrate solution. **3.** Lead carbonate **4.** a) Al^{3+} b) O^{2-} c) Ca^{2+} d) Li^+ e) F^- f) Ni^{2+} g) Ti^{3+} h) SO_4^{2-} i) OH^- j) CO_3^{2-} k) NO_3^- l) CH_3COO^- m) PO_4^{3-}

Ionic equations

Formulae of ionic compounds

When writing formulae for ionic compounds the number of positive charges must be exactly balanced by the number of negative charges. For example, the ionic formula of magnesium oxide is $Mg^{2+}O^{2-}$. The 2+ charge on the magnesium ion is exactly balanced by the 2− charge on the oxide ion.

To work out the ionic formula for aluminium oxide:
The aluminium ion has charge 3+ and the oxide ion has charge 2−. To balance the charges, two aluminium ions are needed for every three oxide ions, so aluminium oxide has the ionic formula $(Al^{3+})_2(O^{2-})_3$. (When the number of ions is greater than one, the formula of the ion should be enclosed in brackets.)

Likewise the formula of ammonium sulphate is $(NH_4^+)_2SO_4^{2-}$ since the ammonium ion is NH_4^+ and the sulphate ion is SO_4^{2-}.

What is a spectator ion?

A spectator ion is an ion which is present during a chemical reaction but does not change or take part in the reaction. Since they are present on both sides of the ionic equation, they are easily identified as the following examples show.

Writing ionic equations without spectator ions

Acid reacting with an alkali

In aqueous solution dilute acids contain hydrogen ions and other ions. For example, dilute hydrochloric acid contains hydrogen ions, $H^+(aq)$ and chloride ions, $Cl^-(aq)$. The ionic equation for hydrochloric acid reacting with sodium hydroxide solution is: $H^+Cl^-(aq) + Na^+OH^-(aq) \rightarrow Na^+Cl^-(aq) + H_2O(l)$

The sodium ions and chloride ions are spectator ions.

Rewriting the equation without spectator ions gives:
$H^+(aq) + OH^-(aq) \rightarrow H_2O(l)$

This equation is valid for all acid/alkali reactions.

Acid reacting with a metal oxide

The ionic equation for nitric acid reacting with solid copper(II) oxide is:
$2H^+NO_3^-(aq) + Cu^{2+}O^{2-}(s) \rightarrow Cu^{2+}(NO_3^-)_2(aq) + H_2O(l)$

The spectator ions in this reaction are copper(II) ions and nitrate ions.

Rewriting the equation without spectator ions gives:
$2H^+(aq) + O^{2-}(s) \rightarrow H_2O(l)$

This equation is valid for all acid/metal oxide reactions.

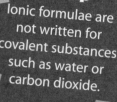

Top Tip
Ionic formulae are not written for covalent substances such as water or carbon dioxide.

Acid reacting with a metal carbonate

The ionic equation for hydrochloric acid reacting with solid calcium carbonate is:

$2H^+Cl^-(aq) + Ca^{2+}CO_3^{2-}(s) \rightarrow Ca^{2+}(Cl^-)_2(aq) + H_2O(l) + CO_2(g)$

The calcium ions and chloride ions are the spectator ions. Rewriting the equation without spectator ions gives:

$2H^+(aq) + CO_3^{2-}(s) \rightarrow H_2O(l) + CO_2(g)$

This equation is valid for all acid/metal carbonate reactions.

Precipitation

The ionic equation for sodium carbonate solution reacting with copper(II) sulphate solution is:

$(Na^+)_2CO_3^{2-}(aq) + Cu^{2+}SO_4^{2-}(aq) \rightarrow Cu^{2+}CO_3^{2-}(s) + (Na^+)_2SO_4^{2-}(aq)$

The copper(II) ions have reacted with the carbonate ions to form the solid precipitate copper(II) carbonate and so the spectator ions are the sodium and sulphate ions. Rewriting the equation without spectator ions gives:

$Cu^{2+}(aq) + CO_3^{2-}(aq) \rightarrow Cu^{2+}CO_3^{2-}(s)$

Quick Test

1. Write ionic formulae for

a) sodium oxide **b)** lead(II) iodide **c)** magnesium chloride **d)** calcium oxide

e) calcium chloride **f)** calcium hydroxide **g)** ammonium nitrate

h) ammonium carbonate

2. What is meant by spectator ions?

3. Name the spectator ions in the following reactions

a) $H^+NO_3^-(aq) + K^+OH^-(aq) \rightarrow K^+NO_3^-(aq) + H_2O(l)$

b) $2H^+Cl^-(aq) + Mg^{2+}CO_3^{2-}(s) \rightarrow Mg^{2+}(Cl^-)_2(aq) + H_2O(l)$

c) $2H^+Cl^-(aq) + Zn^{2+}O^{2-}(s) \rightarrow Zn^{2+}(Cl^-)_2(aq) + H_2O(l)$

d) $(NH_4^+)_2CO_3^{2-}(aq) + Mg^{2+}SO_4^{2-}(aq) \rightarrow Mg^{2+}CO_3^{2-}(s) + (NH_4^+)_2SO_4^{2-}(aq)$

Answers 1. a) $(Na^+)_2O^{2-}$ **b)** $Pb^{2+}(I^-)_2$ **c)** $Mg^{2+}(Cl^-)_2$ **d)** $(Cl^-)_2$ **e)** $Ca^{2+}O^{2-}$ **f)** $Ca^{2+}(OH^-)_2$ **g)** $NH_4^+NO_3^-$ **h)** $(NH_4^+)_2CO_3^{2-}$

2. Spectator ions are present during the reaction but do not take part or change during the reaction.

3. a) potassium and nitrate ions **b)** magnesium and chloride ions **c)** zinc and chloride ions **d)** ammonium and sulphate ions

The electrochemical series

Batteries and simple cells

Batteries provide us with a convenient and portable source of electricity. The **electricity** is generated when the chemicals they contain, react with each other. Strictly speaking, a device that generates electricity from a chemical reaction is a **cell**, and a battery is a collection of two or more cells joined together.

A simple cell, like the one shown below, consists of two different metals dipping into an **electrolyte**.

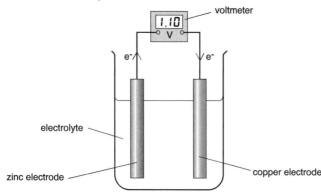

A reading is obtained on the **voltmeter** proving that electricity is produced. The **current** of electricity is carried by **electrons** (e^-) from the zinc electrode through the connecting wires to the copper electrode. The electrons are formed at the zinc electrode when the zinc atoms ionise:

$$Zn(s) \rightarrow Zn^{2+}(aq) + 2e^-$$

The electrolyte is a solution of an ionic compound and its purpose is to complete the circuit. The current is carried through the electrolyte by ions.

When different metal pairs are used, it is found they have different cell voltages. This leads to the electrochemical series. It can be found on page 7 of your Data Booklet.

The electrochemical series is useful in making predictions about cells:

- Electrons will flow in the external circuit of a cell from the metal which is higher in the electrochemical series to the one that is lower.
- The further apart the two metals are in the electrochemical series, the larger is the cell voltage.

Top Tip
In a cell, the current of electricity is carried by electrons through the connecting wires and by ions through the electrolyte.

Ion-bridge cells

An example of an **ion-bridge cell** is shown at the top of the opposite page.

It is made up of two half cells, each consisting of a metal dipping into a solution containing ions of the same metal. The **ion bridge** (or salt bridge) can be a piece of filter paper soaked in an electrolyte. Its purpose is to complete the circuit by allowing ions to flow from one half cell to the other. Since magnesium is higher in the electrochemical series than copper, electrons will flow from magnesium to copper.

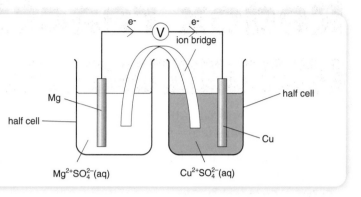

It is possible to set up ion-bridge cells in which one or both half cells involve a non-metal rather than a metal. In non-metal half cells, carbon rods are normally used as electrodes. Carbon is suitable because it conducts electricity.

Displacement reactions

When copper is added to silver(I) nitrate solution, silver forms on the surface of the copper and the solution turns blue.

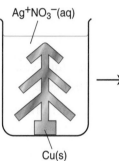

$Cu(s) + 2Ag^+NO_3^-(aq) \rightarrow Cu^{2+}(NO_3^-)_2(aq) + 2Ag(s)$
(brown) (colourless)   (blue)   (grey)

The copper has **displaced** the silver from the silver(I) nitrate solution and the reaction can be described as a **displacement reaction**. The reaction takes place because copper is higher than silver in the electrochemical series. In general, a displacement reaction occurs when a metal is added to a solution containing ions of a metal lower in the electrochemical series.

Displacement reactions can be used to establish the position of hydrogen in the electrochemical series. Certain metals react with acids to produce hydrogen gas, e.g.

$Mg(s) + 2H^+Cl^-(aq) \rightarrow Mg^{2+}(Cl^-)_2(aq) + H_2(g)$

Magnesium has displaced hydrogen from the acid. All metals above copper in the electrochemical series displace hydrogen from acids, but those below, do not. This implies that hydrogen occupies a position in the electrochemical series between lead and copper.

Top Tip
Remember that a displacement reaction occurs when a metal is added to a solution containing ions of a species lower in the electrochemical series.

Quick Test

1. What is the purpose of an electrolyte in a cell?

2. In which direction will electrons flow in the external circuit of a Zn/Mg cell?

3. Which of the following cells would have the highest voltage?
   a) Ag/Pb   **b)** Zn/Pb   **c)** Mg/Ag   **d)** Zn/Mg

4. Name the type of particles that carry the current of electricity through an electrolyte.

5. In which of the following will a displacement reaction take place?
   a) $Pb(s) + Fe^{2+}SO_4^{2-}(aq)$ **b)** $Zn(s) + Mg^{2+}SO_4^{2-}(aq)$
   c) $Mg(s) + Cu^{2+}SO_4^{2-}(aq)$ **d)** $Ni(s) + Na^+Cl^-(aq)$

Answers 1. To complete the circuit. **2.** from Mg to Zn. **3.** C. **4.** Ions. **5.** C.

Redox reactions

Oxidation and reduction reactions

Let's look at the displacement reaction described on the previous page:

$$Cu(s) + 2Ag^+NO_3^-(aq) \rightarrow Cu^{2+}(NO_3^-)_2(aq) + 2Ag(s)$$

The copper atoms have been converted into copper(II) ions. To do this, they each have to lose two electrons:

$$Cu(s) \rightarrow Cu^{2+}(aq) + 2e^-$$

The process in which a reactant loses electrons is called **oxidation** and so the copper atoms have been oxidised. An equation like the one immediately above is called an **ion-electron equation** because it contains both ions and electrons.

The electrons that the copper atoms lose are gained by the silver(I) ions when they form silver atoms:

$$Ag^+(aq) + e^- \rightarrow Ag(s)$$

The process in which a reactant gains electrons is known as **reduction** and so the silver(I) ions have been reduced.

Oxidation and reduction reactions always occur together. These **red**uction/**ox**idation reactions are known as **redox** reactions. Equations for redox reactions can be obtained by combining the ion-electron equations for the oxidation and reduction processes. In our example, you will notice that the oxidation equation contains two electrons, while the reduction equation contains only one. Before adding them together, we multiply the reduction equation by two in order that the redox equation is balanced and the electrons cancel out:

$Cu(s) \rightarrow Cu^{2+}(aq) + 2e^-$	**Oxidation 1**
$2Ag^+(aq) + 2e^- \rightarrow 2Ag(s)$	**Reduction 2**

Adding 1 and 2: $Cu(s) + 2Ag^+(aq) \rightarrow Cu^{2+}(aq) + 2Ag(s)$ **Redox**

Notice that the redox equation does not contain spectator ions. In general, redox equations must not contain electrons or spectator ions.

Ion-electron equations can be found on page 7 of your Data Booklet. They are all written as reductions. To write an oxidation ion-electron equation, you must find the relevant equation on page 7 and reverse it.

Examples of oxidation and reduction reactions

A metal reacting to form a metal compound is an oxidation reaction. For example, when magnesium burns, it forms magnesium oxide:

$$2Mg(s) + O_2(g) \rightarrow 2Mg^{2+}O^{2-}(s)$$

The magnesium atoms have been converted into magnesium ions and, to do this, they must lose electrons. The magnesium atoms have therefore been oxidised.

A metal compound reacting to form a metal is a reduction reaction. For example, when mercury(II) oxide is heated it decomposes to form mercury:

$$2Hg^{2+}O^{2-}(s) \rightarrow 2Hg(l) + O_2(g)$$

The mercury(II) ions have been converted into mercury atoms and, to do this, they must gain electrons. The mercury(II) ions have therefore been reduced.

Oxidation and reduction in electrolysis

When a solution or melt of an ionic compound is electrolysed, the compound may break up into its elements. Consider the electrolysis of copper(II) chloride solution.

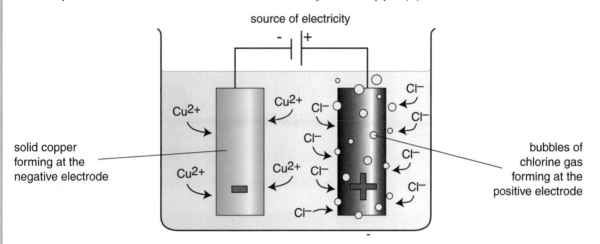

source of electricity

solid copper forming at the negative electrode

bubbles of chlorine gas forming at the positive electrode

The positive copper(II) ions are attracted to the negative electrode where they gain electrons and are reduced to copper atoms:
$$Cu^{2+}(aq) + 2e^- \rightarrow Cu(s) \quad \text{Reduction}$$

The negative chloride ions are attracted to the positive electrode where they must be oxidised. To get the ion-electron equation for this oxidation we look up the relevant equation on page 7 of the Data Booklet and reverse it:
$$2Cl^-(aq) \rightarrow Cl_2(g) + 2e^- \quad \text{Oxidation}$$

Notice that oxidation occurs at the positive electrode and reduction occurs at the negative electrode. This is always the case no matter which ionic compound is electrolysed.

Quick Test

1. What is meant by oxidation?

2. Write an ion-electron equation for:
   a) the reduction of bromine, Br_2 **b)** the oxidation of aluminium, Al.

3. In the reaction, $Zn(s) + 2H^+Cl^-(aq) \rightarrow H_2(g) + Zn^{2+}(Cl^-)_2(aq)$, which of the following is reduced?   **A** $Zn(s)$ **B** $H^+(aq)$ **C** $Cl^-(aq)$ **D** $H_2(g)$

4. Given the following ion-electron equations:
   $$Fe^{2+}(aq) \rightarrow Fe^{3+}(aq) + e^- \qquad Cl_2(aq) + 2e^- \rightarrow 2Cl^-(aq),$$
   write the equation for the redox reaction.

5. Write the ion-electron equation for the reaction that takes place at the negative electrode during the electrolysis of molten sodium iodide.

Answers 1. A process in which a reactant loses electrons. **2. a)** $Br_2(l)+2e^- \rightarrow 2Br^-(aq)$ **b)** $Al(s) \rightarrow Al^{3+}(aq) + 3e^-$ **3.** B **4.** $2Fe^{2+}(aq) + Cl_2(aq) \rightarrow 2Fe^{3+}(aq) + 2Cl^-(aq)$
5. $Na^+ + e^- \rightarrow Na$

Reactions of metals

Reaction with oxygen

Metals like potassium, sodium and lithium react so readily with oxygen in the air that they have to be stored under oil. They are very **reactive** metals. Gold, on the other hand, does not require to be stored in any special way because it does not react with oxygen. It is an **unreactive** metal.

The reactions of metals with oxygen can be demonstrated using the apparatus shown. When the potassium permanganate is heated, it produces oxygen, which then reacts with the hot metal.

From the speed of burning and colour of the flame, the metals tested can be placed in order of reactivity. The most reactive metal would burn the fastest and with the brightest flame.

In general, when a metal reacts with oxygen, the corresponding metal oxide is formed. For example:

$2Cu(s) + O_2(g) \rightarrow 2Cu^{2+}O^{2-}(s)$

potassium permanganate / plug of mineral wool / metal / O_2 / HEAT / HEAT

Reaction with water

We know that the Group 1 metals lithium, sodium and potassium react vigorously with water. The potassium/water reaction is the fastest and the lithium/water reaction is the slowest. The Group 2 metal calcium reacts with water, but not as vigorously as the Group 1 metals. Magnesium, another Group 2 metal, reacts only very slowly with water. These observations give us the following order of reactivity, starting with the most reactive: potassium, sodium, lithium, calcium, magnesium.

In general, when a metal reacts with water, hydrogen gas and the corresponding metal hydroxide solution are formed. For example:

$Ca(s) + 2H_2O(l) \rightarrow H_2(g) + Ca^{2+}(OH^-)_2(aq)$

When a metal reacts with a dilute acid, bubbles of hydrogen gas are given off. The speed at which the bubbles are produced gives us some idea of the reactivity of metals; the greater the speed the more reactive the metal.

Only metals above hydrogen in the electrochemical series react with a dilute acid. Those metals below hydrogen, like copper, mercury, silver and gold, do not react with a dilute acid.

In general, when a metal reacts with a dilute acid, hydrogen gas and the corresponding metal salt are formed. For example:

$Mg(s) + (H^+)_2SO_4^{2-}(aq) \rightarrow H_2(g) + Mg^{2+}SO_4^{2-}(aq)$

Reactivity series

By taking a wide variety of metals and looking at their reaction rates with oxygen, water and dilute acid, a longer list of metals, arranged in order of reactivity, can be obtained. This list is shown in the first column of the table, and as we go down the column, metal reactivity decreases. A summary of the reactions of metals with oxygen, water and dilute acid is also given in the table.

Metal	Reaction with oxygen	Reaction with water	Reaction with dilute acid
potassium	*reacts vigorously*	*reacts vigorously*	*reacts explosively*
sodium		**metal + water**	
lithium		↓	
calcium		**hydrogen + metal hydroxide**	
magnesium	**metal + oxygen**	*reacts very slowly*	**metal + dilute acid**
aluminium	↓		↓
	metal oxide		**hydrogen + metal salt**
zinc			
iron			
tin		**no reaction with water**	
lead			*reacts very slowly*
copper	*reacts very slowly*		
mercury			**no reaction with dilute acid**
silver	**no reaction with oxygen**		
gold			

Notice that the metal reactivity series is very similar, although not identical, to the electrochemical series given on page 7 of your Data Booklet. The electrochemical series is therefore a good guide to metal reactivity.

Quick Test

1. Write a balanced equation for the reaction between aluminium and oxygen.

2. Write a balanced equation for the reaction between sodium and water.

3. Write a balanced equation for the reaction between zinc and dilute hydrochloric acid.

4. Metal A reacts slowly with water; metal B reacts with oxygen but not with dilute acid; metal C reacts with dilute acid but not with water.

   a) Using the above information, list metals A, B and C in order of reactivity starting with the most reactive.

   b) Suggest a name for metal B.

Answers 1. $4Al(s) + 3O_2(g) \rightarrow 2(Al^{3+})_2(O^{2-})_3(s)$. **2.** $2Na(s) + 2H_2O(l) \rightarrow H_2(g) + 2Na^+OH^-(aq)$. **3.** $Zn(s) + 2H^+Cl^-(aq) \rightarrow H_2(g) + Zn^{2+}(Cl^-)_2(aq)$ **4. a)** ACB **b)** Copper

Metal ores

Extraction of metals

Only a few metals, such as gold and silver, are found in the Earth's crust in an **uncombined state**, i.e. they are present as atoms and not as ions. The vast majority of metals are found in the **combined state**, i.e. they exist as compounds and so the metal is present as ions. These naturally occurring compounds of metals are known as **ores**.

Extracting a metal from its ore is an example of a **reduction reaction** because the metal ions in the ore have to gain electrons in order to form metal atoms.

The amount of energy needed to reduce metal ions to metal atoms, i.e. to extract a metal from its ore, depends on the reactivity of the metal. The three main methods of extracting metals are outlined below:

Extracting metals by heat alone

Very little energy is needed to extract unreactive metals such as mercury, silver and gold. They can be extracted simply by heating their metal oxides. For example: $2Hg^{2+}O^{2-}(s) \rightarrow 2Hg(l) + O_2(g)$

Extracting metals by heating with carbon or carbon monoxide

The oxides of metals more reactive than mercury do not decompose on heating. Heat alone provides insufficient energy. So the oxides of metals in the middle region of the reactivity series have to be heated with carbon or carbon monoxide. For example:
$2Pb^{2+}O^{2-}(s) + C(s) \rightarrow 2Pb(s) + CO_2(g)$
$Pb^{2+}O^{2-}(s) + CO(g) \rightarrow Pb(s) + CO_2(g)$

Extracting metals by electrolysis

For metals above zinc in the reactivity series, even more energy is needed to extract them. This can be achieved by using electricity. For example, aluminium is extracted from aluminium oxide (bauxite) by electrolysing the liquid ore. The aluminium ions are reduced to aluminium atoms at the negative electrode: $Al^{3+}(l) + 3e^- \rightarrow Al(s)$

The other highly reactive metals are obtained by electrolysing their molten chlorides rather than their molten oxides.

The methods used in extracting metals from their ores are summarised below:

Metal	Method of extraction
potassium	
sodium	
lithium	Electrolysis of
calcium	molten ore
magnesium	
aluminium	

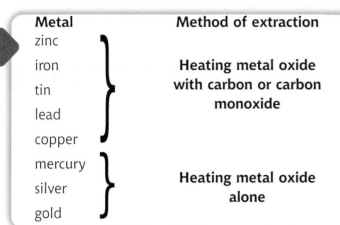

Metal	Method of extraction
zinc	
iron	
tin	Heating metal oxide with carbon or carbon monoxide
lead	
copper	
mercury	
silver	Heating metal oxide alone
gold	

Top Tip
By considering the position of a metal in the reactivity series you must be able to suggest the method used to extract it from its ore. You can use the electrochemical series on page 7 of your Data Booklet as a guide to metal reactivity.

Extraction of iron – the blast furnace

Iron is produced from **iron ore** in the **blast furnace**, a diagram of which is shown opposite. Iron ore (iron(III) oxide) is loaded in at the top along with coke (carbon) and limestone. Hot air is blown in at the base and it reacts with the carbon to form carbon dioxide:

$C(s) + O_2(g) \rightarrow CO_2(g)$

As the carbon dioxide rises up the tower it reacts with more carbon to form carbon monoxide:

$CO_2(g) + C(s) \rightarrow 2CO(g)$

The carbon monoxide then reduces the iron(III) oxide to iron:

$(Fe^{3+})_2(O^{2-})_3(s) + 3CO(g)$
$\downarrow$
$2Fe(l) + 3CO_2(g)$

The molten iron runs down the tower where it is tapped off at the bottom.

Top Tip
Make sure you know the two reactions involved in producing carbon monoxide in the blast furnace.

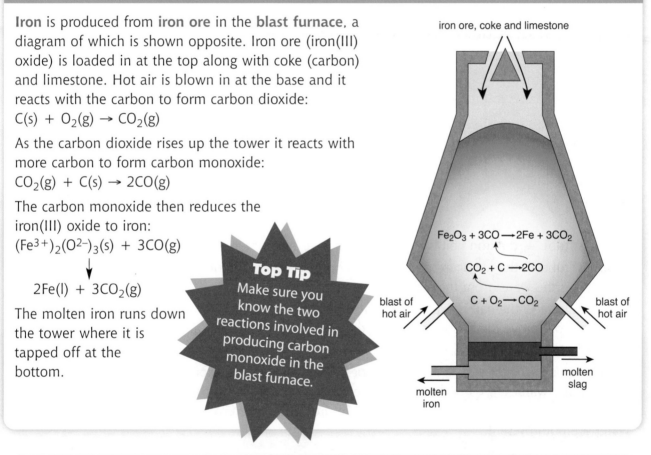

iron ore, coke and limestone

$Fe_2O_3 + 3CO \longrightarrow 2Fe + 3CO_2$

$CO_2 + C \longrightarrow 2CO$

$C + O_2 \longrightarrow CO_2$

blast of hot air

blast of hot air

molten slag

molten iron

Quick Test

1. Write a balanced equation for the reaction that takes place when silver(I) oxide is heated.

2. Write a balanced equation for the reaction that takes place when copper(II) oxide is heated with carbon monoxide.

3. Which method of metal extraction would be used to obtain sodium from sodium chloride?

4. Write balanced equations for the two reactions involved in producing carbon monoxide in the blast furnace.

Answers 1. $2(Ag^+)_2O^{2-}(s) \rightarrow 4Ag(s) + O_2(g)$ **2.** $Cu^{2+}O^{2-}(s) + CO(g) \rightarrow Cu(s) + CO_2(g)$ **3.** Electrolyse molten sodium chloride. **4.** $C(s) + O_2(g) \rightarrow CO_2(g)$ followed by $CO_2(g) + C(s) \rightarrow 2CO(g)$

Corrosion

What is corrosion?

Corrosion is a chemical reaction that takes place on the surface of a metal. The metal reacts with substances in the atmosphere to form a metal compound. Corrosion, therefore, is an example of **oxidation** since the metal atoms must lose electrons when they form the metal ions present in the compound.

Different metals corrode at different rates. The rate of corrosion of a metal is related to its reactivity; the more reactive the metal, the quicker it will corrode.

Conditions required for rusting

The corrosion of iron is usually called **rusting**. The experiment shown below can be set up to show which substances in the atmosphere cause iron to rust.

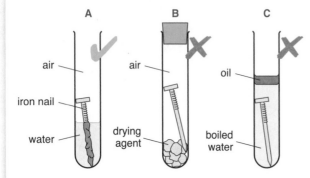

Top Tip
Remember that both oxygen and water must be present for iron to rust.

- In test tube **A**, the iron nail is exposed to both air and water.
- In test tube **B**, the nail is exposed to air but not to water. The drying agent removes the water from the air.
- In test tube **C**, the nail is exposed to water but not to air. The water has been boiled to remove the dissolved air and the oil layer prevents any air from re-dissolving.

After a few days, rust forms only in test tube **A** where the iron is exposed to both air and water. Air contains several gases, but it can be shown that when iron rusts, the gas that is used up is oxygen. So, iron rusts when oxygen and water are both present.

Factors that affect the rate of rusting

The rate at which iron rusts is affected by changes in temperature and by the presence of salt and acid rain.

- Temperature: An increase in temperature will increase the rate of rusting, e.g. a car exhaust rusts more rapidly than the rest of the bodywork because of its higher temperature.
- Salt and acid rain: Both salt and acid rain increase the rate of rusting, e.g. salt spread on roads during winter increases the rate of corrosion of cars.

Rusting – a redox reaction

When iron rusts, the iron atoms are oxidised to iron(II) ions:

$Fe(s) \rightarrow Fe^{2+}(aq) + 2e^-$ Oxidation

These iron(II) ions are then further oxidised to iron(III) ions:

$Fe^{2+}(aq) \rightarrow Fe^{3+}(aq) + e^-$ Oxidation

Since water and oxygen have to be present for iron to rust, it must be the water and oxygen molecules that are reduced:

$2H_2O(l) + O_2(g) + 4e^- \rightarrow 4OH^-(aq)$ Reduction

The iron(III) ions and hydroxide ions that are formed in the redox process react to form iron(III) hydroxide which is present in rust:

$Fe^{3+}(aq) + 3OH^-(aq) \rightarrow Fe^{3+}(OH^-)_3(s)$

In effect, small cells are set up on the surface of the iron in which electrons are transferred from the iron to the water and oxygen. This current of electricity is carried by ions, and since water contains so few ions, rusting is a slow process. In the presence of an electrolyte, however, rusting is accelerated. The reason is that an electrolyte contains a much higher concentration of ions than water, and is a much better conductor of electricity. This explains why electrolytes, such as salt solution and acid rain, speed up rusting.

Top Tip

Remember that ferroxyl indicator turns blue in the presence of $Fe^{2+}(aq)$ ions and pink in the presence of OH^- ions.

Ferroxyl indicator

Ferroxyl indicator can be used to detect rusting. When iron rusts, iron(II) ions are formed initially and ferroxyl indicator turns from yellow to blue in the presence of $Fe^{2+}(aq)$ ions.

During rusting, hydroxide ions are also produced. Ferroxyl indicator turns from yellow to pink in the presence of $OH^-(aq)$ ions.

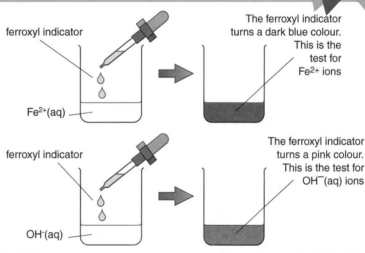

ferroxyl indicator

$Fe^{2+}(aq)$

The ferroxyl indicator turns a dark blue colour. This is the test for Fe^{2+} ions

ferroxyl indicator

$OH^-(aq)$

The ferroxyl indicator turns a pink colour. This is the test for $OH^-(aq)$ ions

Quick Test

1. Which of the following metals will corrode at the fastest rate?
   A copper   **B** magnesium   **C** silver   **D** iron

2. Which two substances must be present for iron to rust?

3. Write ion-electron equations for the two oxidation reactions that take place when iron rusts.

4. Why does salt solution accelerate rusting?

5. Name the indicator used to detect rusting.

More on corrosion

The effect of other metals on rusting

The following experiment can be set up to demonstrate what effect attaching different metals to iron has on rusting. The iron nail in test tube **A** has no metal attached to it and is the control. In **B**, copper, a metal lower than iron in the electrochemical series, is attached to the nail. In **C**, magnesium, which is higher than iron in the electrochemical series, is attached. In all three test tubes, ferroxyl indicator is present to show if rusting occurs.

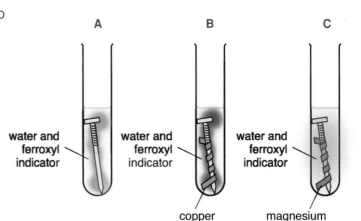

The blue colour in **A** and **B** indicates the presence of $Fe^{2+}(aq)$ ions, i.e. the iron nails have rusted. Since the blue colour in **B** is more extensive, this means that attaching copper to iron speeds up rusting. No blue colour appears in **C** – this means that attaching magnesium to iron prevents rusting. The pink colour in **C** indicates the presence of $OH^-(aq)$ ions. They are formed when the magnesium reacts with water.

In general, when iron is attached to a metal lower in the electrochemical series, it rusts more quickly but when attached to a metal higher in the electrochemical series, it does not rust.

Top Tip
When attached to iron, metals higher in the electrochemical series will prevent rusting but lower metals will accelerate it.

Rust prevention – physical protection

If iron is to rust, it must be in contact with oxygen and water. If a coating is put on the surface of the iron, it will act as a barrier to the oxygen and water and prevent rusting. This is called **physical protection**. Examples include:

- **Painting**: car bodies, iron gates and bridges (Forth Rail Bridge)
- **Oiling and greasing**: moving parts in industrial and agricultural machinery, bike chains
- **Plastic coating**: garden wires and fences, fridge shelves
- **Tin-plating**: food cans (tin cans). Traditionally tin cans were made of iron on which there was a thin coating of tin. If the tin coating is scratched, the iron will rust even more rapidly.
- **Galvanising**: car exhausts and nails. Galvanised iron is iron coated with zinc. This is done by dipping the iron into molten zinc.
- **Electroplating**: silver-plated cutlery, chromium-plated car bumpers. Electroplated iron has a thin layer of another metal coated on it. This can be achieved by electrolysis.

Rust prevention – electrochemical protection

When iron rusts, the iron atoms lose electrons. So, if electrons are pushed back onto the iron, then it will not rust. This is the basis of **electrochemical protection**. Examples include:

- **Sacrificial protection**

  This involves linking or attaching a metal higher than iron in the electrochemical series to the iron structure. This metal will supply the iron with electrons thus preventing it from rusting. The attached metal, however, will corrode more rapidly than it would on its own and so is being sacrificed to protect the iron. Hulls of ships and oil rigs are protected sacrificially by attaching blocks of zinc to them.

- **Using electricity**

  This involves connecting the iron to the negative terminal of a battery. It supplies electrons to the iron thus preventing it from rusting. This method is used in cars when the negative terminal of the battery is connected to the iron bodywork.

- **Galvanising**

  Galvanised iron is iron coated with zinc. At first, the zinc offers the iron physical protection. If the zinc coating is broken, however, the zinc can then offer the iron electrochemical protection. The reason is that zinc is higher than iron in the electrochemical series and supplies electrons to the iron thus preventing it from rusting.

Top Tip
Only metals higher than iron in the electrochemical series can offer iron sacrificial protection.

Quick Test

1. The following cell was set up:

   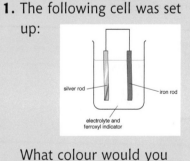

   silver rod — iron rod
   electrolyte and ferroxyl indicator

   What colour would you expect to see around the iron rod?

2. Iron wires coated in plastic do not rust. Explain how the plastic coating prevents the iron from rusting.

3. a) What name is given to iron coated with zinc?

   b) Why doesn't the iron rust even when the zinc coating is scratched?

4. When magnesium is attached to iron, it provides sacrificial protection to the iron. Explain how the magnesium protects the iron.

Answers 1. Blue **2.** The plastic coating puts a barrier between the iron and the oxygen and water **3. a)** Galvanised iron **b)** Because zinc is higher than iron in the electrochemical series **4.** Magnesium is higher than iron in the electrochemical series and supplies the iron with electrons

PPA 1: Preparation of a salt

Introduction

A salt is formed when the hydrogen ions of an acid are replaced by metal ions or ammonium ions. For example, replacing the hydrogen ions in sulphuric acid by magnesium ions produces the salt, magnesium sulphate. Magnesium sulphate can be made by reacting excess magnesium or magnesium carbonate with sulphuric acid. The fact that a gas is produced on reacting magnesium or magnesium carbonate with an acid allows us to tell when the reaction is complete. When the reaction is complete, no more bubbles of gas will appear, and the excess magnesium or magnesium carbonate remains as a solid in the reaction mixture.

Aim

To prepare a pure sample of magnesium sulphate, by reacting magnesium or magnesium carbonate with dilute sulphuric acid.

Procedure

20 cm³ of dilute sulphuric acid was measured into a small beaker. Magnesium or magnesium carbonate was added in excess while stirring. The reaction mixture was filtered, and the filtrate (magnesium sulphate solution) was transferred to an evaporating dish. The magnesium sulphate solution was heated until about half the water remained. The evaporating dish was allowed to cool slowly to let crystals form. The three steps involved in preparing magnesium sulphate are shown in the diagram below.

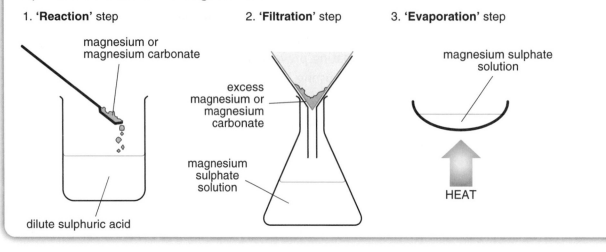

1. **'Reaction' step**
magnesium or magnesium carbonate
dilute sulphuric acid

2. **'Filtration' step**
excess magnesium or magnesium carbonate
magnesium sulphate solution

3. **'Evaporation' step**
magnesium sulphate solution
HEAT

Results

Colourless magnesium sulphate crystals formed on cooling. They had the following shape:

Conclusion

Crystals of magnesium sulphate were prepared. The word equations for the reactions are:

magnesium + sulphuric acid → magnesium sulphate + hydrogen gas

magnesium carbonate + sulphuric acid → magnesium sulphate + water + carbon dioxide gas

Points to note

- An excess of magnesium or magnesium carbonate had to be used to make sure all the acid was used up, otherwise, the salt would be contaminated by the acid.
- Since magnesium and hydrogen gas are both highly flammable, all ignition sources, like lighted Bunsen burners, must be absent.

Quick Test

1. Why was the magnesium or magnesium carbonate added in excess to the sulphuric acid?

2. How can you tell that the reaction is complete?

3. Write the word equation for magnesium carbonate reacting with sulphuric acid.

4. What is the name of the filtrate in the filtration step?

5. Which gas is produced when magnesium reacts with sulphuric acid?

6. Why should the evaporating basin be allowed to cool slowly?

Answers 1. To make sure that all the acid is used up. **2.** When no more bubbles of gas are produced and some unreacted Mg or MgCO₃ is left in the bottom of the beaker. **3.** Magnesium carbonate + sulphuric acid → magnesium sulphate + water + carbon dioxide gas **4.** Magnesium sulphate solution. **5.** Hydrogen. **6.** To let crystals form.

PPA 2: Factors affecting voltage

Introduction

A cell is a device in which a chemical reaction is used to produce electricity. A simple cell can be made by dipping two different metal electrodes into a solution that is able to conduct a current of electricity, i.e. an electrolyte. The voltage that the cell generates can be measured using a voltmeter.

Aim

To find out if changing the metals or changing the electrolyte in a cell affects its voltage.

Procedure

The metal electrodes were cleaned using emery paper and then washed. In each experiment the apparatus shown opposite was used. In the 'changing the metals' experiment, three cells: zinc/copper, copper/iron and zinc/iron were set up with sodium chloride solution as the electrolyte. In the 'changing the electrolyte' experiment, copper and zinc were used as electrodes and the three electrolytes tested were: sodium chloride, hydrochloric acid and sodium hydroxide solutions. For each cell, the voltage was measured and then re-measured after removing the electrodes and re-inserting them into the electrolyte.

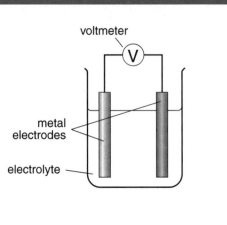

Results for 'changing the metals'

Metals used	Voltage 1 (V)	Voltage 2 (V)	Average voltage (V)
Zinc and copper	0·75	0·71	0·73
Copper and iron	0·29	0·29	0·29
Zinc and iron	0·33	0·31	0·32

Conclusion

Changing the metals used in a cell affects its voltage.

Results for 'changing the electrolyte"

Electrolyte solution	Voltage 1 (V)	Voltage 2 (V)	Average voltage (V)
Sodium chloride	0·14	0·14	0·14
Hydrochloric acid	0·60	0·62	0·61
Sodium hydroxide	0·25	0·29	0·27

Conclusion

Changing the electrolyte in a cell affects its voltage.

Points to note

- In both experiments, the factors that were kept the same included temperature, concentration of electrolyte, volume of electrolyte and the distance between the electrodes. In the 'changing the metals' experiment, the same electrolyte was used and in the 'changing the electrolyte' experiment, the same metals were used as electrodes.
- The electrodes were initially cleaned with emery paper and then washed in order to improve the contact between them and the electrolyte.
- Duplicate voltages were obtained for each cell to show that the results were valid and reliable.

Quick Test

1. What produces the electricity in a cell?
2. How can a simple cell be made?
3. Why were the metal electrodes cleaned with emery paper?
4. Why were duplicate voltages obtained?
5. Which factors were kept the same in all the experiments?
6. What was the electrolyte in the 'changing the metals' experiment?
7. Which electrodes were used in the 'changing the electrolyte' experiment?

Answers 1. A chemical reaction. **2.** By dipping two different metals into an electrolyte. **3.** To improve the contact between them and the electrolyte. **4.** To ensure the results were valid and reliable. **5.** Temperature, concentration, volume of the electrolyte and distance between the electrodes. **6.** Sodium chloride solution. **7.** Copper and zinc.

PPA 3: Reaction of metals with oxygen

Introduction

Some metals, like potassium and sodium, are highly reactive but others, like platinum and gold, are unreactive. The majority of metals lie between these two extremes. One way of placing metals in order of reactivity is to compare their reactions with oxygen. Potassium permanganate is used to provide the oxygen.

Aim

To place, zinc, copper and magnesium in order of reactivity by observing the ease with which they react with oxygen.

Procedure

Using the apparatus shown below, the zinc was first heated until it was red hot and then the heat was transferred to the potassium permanganate so that the oxygen produced passed over the hot metal. Alternate heating of the zinc and potassium permanganate was continued until the reaction was complete. The experiment was repeated with copper and then with magnesium.

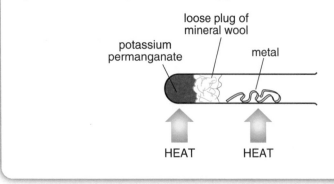

Results

Metal	Observations
Zinc	Reasonably fast reaction; yellow glow
Copper	Slow reaction; dull red glow
Magnesium	Fast reaction; very bright white light

Conclusion

The order of reactivity of the metals, from the most reactive to the least reactive, was magnesium, zinc and copper.

Points to note

- The potassium permanganate is used to provide oxygen. It does this when it decomposes on heating.
- The mouth of the test tube must not point at anyone in case a reaction becomes violent and the hot contents of the test tube are ejected.
- To prevent damage to the eyes, they must be shaded from the dazzling white light produced when magnesium burns.
- Magnesium is highly flammable, and apart from when it is being heated in the test tube, naked flames must be kept away from the magnesium.
- Mineral wool irritates the skin and tongs must be used when handling it.

Quick Test

1. How is the oxygen produced in the experiment?
2. What is the heating procedure in the experiment?
3. Which of the three metals tested produced the fastest reaction?
4. Why is it important not to point the mouth of the test tube at anyone?
5. Why must you not look directly at the magnesium when it burns?

Answers 1. By heating potassium permanganate. **2.** Heat the metal until it is red-hot and then transfer the heat to the potassium permanganate, then heat both parts alternately. **3.** Magnesium. **4.** In case the hot contents of the test tube are ejected out and hit someone. **5.** Because the dazzling white light produced might cause damage to the eyes.

Index

acid rain 89, 106
acids 82–89
addition reactions 56–57, 61
adsorption 14
air pollution 43
alcoholic drinks 60–61
alkali metals 6
alkalis 82–83, 88, 93, 97
alkane isomers 50
alkanes 46–47, 56, 58
alkanoic acids 53, 56
alkanols 52–53, 56
alkene isomers 51
alkenes 48–49, 56–57, 58
amines 68–69, 72
amino acid monomer units 72
ammonia 83
argon 7
atomic numbers 16
atoms 16–17

balanced equations 32–35
bases 87
batteries 98
Benedict's test 71, 80–81
biological catalysts 15
bonding 18–19
bonds 18–19
branched-chain alkanes 47
bromine 7
    solution 56–57, 75

carbohydrates 70–71
carboxyl group 53, 72, 74
catalysts 13–15, 71
catalytic hydration reactions 61
cells 98–99, 112–113
chemical reactions see reactions
chlorine 7
collision theory 12–13
combustion 42–43
compounds 8
concentration 12, 84–85, 90
condensation polymerisation 68–69, 72
condensation reaction 62, 68, 74
corrosion 106–109
covalent bonding 18
covalent molecular substances 20–22, 24
covalent network structures 22
cracking 58–59, 66, 78–79
crude oil 44–45
cycloalkane isomers 51
cycloalkanes 49, 56

decomposition 26
dehydration reactions 61, 62
diatomic molecules 20
digestion 71, 73
displacement reactions 99–100
distillation 60–61

electricity 98–99
electrochemical series 98–99
electrolysis 26–27, 40–41
electrolytes 26, 98–99
electrons 98–99
elements 4–7
endothermic reactions 9, 70
enzymes 15, 71, 72
equations, balanced 32–35
esters 53–54, 62–63, 74
ethanol 60–61
exothermic reactions 9, 70

extraction, of metals 104–105

fats 74–75
fermentation 60–61
ferroxyl indicator 107
fluorine 7
formula triangles 34, 84–85
formulae 28–31
fossil fuels 42
fractional distillation 44–45, 58, 66
fractions 44–45
fuels 42–45

gases 4, 7
glycerol 74
graphite 24
groups 5–6, 28

haemoglobin 72
halogens 6
hardening 75
helium 7
heterogeneous catalysts 14, 58
homogeneous catalysts 14
hormones 72
hydration reactions 57, 61, 63
hydrocarbons 42–51
hydrogen ions 83
hydrolysis reactions 63, 71, 73, 75, 80–81
hydroxide ions 83

insoluble salts 94–95
iodine 7
    test 70
ion-bridge cell 98–99
ion-electron equation 100
ion migration 27
ionic bonding 18–19
ionic equations 96–97
ionic lattice 22
ionic substances 24
ions 18–19
iron 105, 107
isomers 50–51
isotopes 17

krypton 7

liquids 4
long-chain polymer molecules 68, 72

mass numbers 16–17
metal carbonates 88, 93, 97
metal oxides 82, 88, 93, 97
metal reactivity 102–103
metallic bonding 19
metallic structures 23
metals 4, 24, 89, 93, 98–106, 114–115
mixtures 8
molecules 20–21
moles 34–35, 84
monomers 66–67

negative ions 19
neon 7
neutralisation 88–89, 90–91
neutrons 16
nitrogen salts 92
noble gases 7, 28
non-metal oxides 82
non-polar covalent bonding 18
nucleus 16

oils 74–75
ores 104–105

oxidation 26, 100–101, 106–107

peptide links 72
Periodic Table 4–5, 28
petrol 58
pH scale 82
photosynthesis 70
plastics 64–69
poisoning 14–15
polar covalent bonding 18
pollutants 43
pollution 65
poly(ethanol) 64
polymerisation 66–67
polymers 66–67
power 90
precipitation 94–95, 97
products 9, 32
proteins 72–73
protons 16

RAM (relative atomic mass) 17
reactants 9, 32
reaction rates 10–15, 36–39
reactions
    of acids 88–89
    addition 56–57, 61
    compound 8
    corrosion 106–107
    cracking 58–59
    endothermic 9
    esterification 62–63
    exothermic 9
    fermentation 60–61
    hydrolysis 63, 71, 73, 75, 80–81
    of metals 102–103
    redox 100–101, 106–107
    reduction 26, 100, 104–105
reactive metals 102, 114–115
redox reactions 100–101, 106–107
reduction reactions 26, 100, 104–105
relative atomic mass (RAM) 17
relative formula mass 34–35
respiration 70
rusting 106–107

salts 88, 92–93, 110–111
saturated compounds 56–57
solids 4
soluble salts 92–93
solutions 8
starches 70, 71
state symbols 32
strong acids 86
strong bases 87
sugars 70
synthetic fibres 64–69

thermoplastics 65
thermosetting plastics 65
titrations 90–91
transition metals 6, 30, 43

units 84
unsaturation 56–57, 76–77

valency 28–31
voltage 98, 112–113
volumetric titration 90–91

weak acids 86
weak bases 87